Sirius Matters

Astrophysics and Space Science Library

Recently Published in the ASSL series

For other titles see www.springer.com/astronomy

Recently Published in the ASSL Series

Sirius Matters

Noah Brosch

Tel Aviv University
Tel Aviv, Israel

 Springer

Dr. Noah Brosch
The Tel Aviv University
Wise Observatory and Dept.
of Astronomy & Astrophysics
School of Physics and Astronomy
69978 Tel Aviv
Israel
noah@wise.tau.ac.il

ISBN 978-1-4020-8318-1   ISBN 978-1-4020-8319-8 (eBook)

Library of Congress Control Number: 2008926593

Cover illustration: A combination of nine frames from the STScI on-line digitized POSS2/UKSTU
IR & POSS2/UKSTU blue images, illustrating the brightness of Sirius. The center of
mosaic is at RA = 6:45:08.9, Dec = −16:42:58. The sky area coverage is $2°.75 \times 2°.73$. Image
processing: Anurag Shevade. Used with permission.

Printed on acid-free paper

9 8 7 6 5 4 3 2 1

springer.com

Contents

List of figures

List of tables

Chapter 1
Introduction

Astronomy is a difficult science and it is a challenge to understand even the stars that compose the brightest system on the night sky: Sirius (α Canis Majoris=α CMa). This is one of the nearest stars and the brightest one as seen from Earth, with the exception of our Sun. At present, perhaps unfortunately, stellar astronomy and in particular the study of the very bright stars are not part of the mainstream astronomy research. The professional attention is now focused on the cutting-edge of cosmology, the early Universe, extremely distant galaxies, the "origins" theme, etc. Bright stars are perceived to be too mundane to be considered an interesting research subject.

Sirius has many unique astronomical aspects that make it worth investigating. It is significantly more massive than the Sun, its outer layers show strange and unusual chemical abundances, it is a member in a double-star system along with one of the most massive white dwarfs in an elliptical orbit, and it is outstandingly luminous in comparison with other nearby stars. The closeness to the Sun implies that Sirius can be studied in significantly more detail than most other stars. In a way, it as much as the Sun may be keys to understand stars, and to understand the immediate Solar neighborhood.

This book attempts to bring some order to the wealth of information about Sirius. The Walker Art Center in Minneapolis carries an interesting writing on its facade: *BITS AND PIECES PUT TOGETHER TO MAKE A SEMBLANCE OF A WHOLE* (see Figure 1.1). This monograph attempts to do something similar for the Sirius binary. Various bits and pieces of information about it, starting with historical and ethnographic information and going to astrophysical models of stellar structure and evolution, have been put together to attempt the derivation of a coherent picture of this fascinating system.

Three main aspects of Sirius are of particular interest: historical and cultural, formative and evolutionary, and those related to the Solar environment; they will be treated more or less in this order in the different chapters of this book. The plan is, therefore, to describe first the ancient observations, bringing the reader up through the ages till the beginning of astronomical photography.

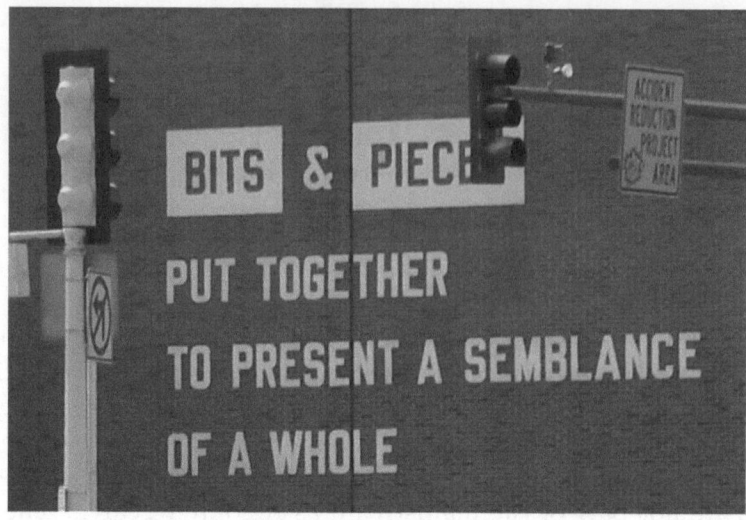

Figure 1.1.   Facade of the Walker Art Center in Minneapolis (Photograph by the author).

Sirius had an important part to play in the history of civilizations, probably because it is so bright compared to the other stars. The historical descriptions and its link to ancient gods will be among the first topics dealt with below. In this context, one must consider also fantastic claims of extraterrestrials visiting the Earth and civilizing its inhabitants; the origin of those ETs is sometimes claimed to be on a planet of the Sirius system.

In this first part of the book (Chapters 2 and 3) I will present controversial evidence for a color change of Sirius in historical times; this has been discussed extensively for more than one century in the astronomical literature, and various explanations have been proposed to account for it, but no explanation has ever been fully accepted by the community. The evidence for the color change itself has been questioned by some authors.

Part two (Chapters 4 and 5) is dedicated to developments since the start of astronomical photography until the beginning of space astronomy. α CMa is a binary star; this was discovered more than a century ago. Its companion, α CMa B, is one of the most massive white dwarfs. This, and its highly elliptical orbit, make Sirius an interesting subject of study in order to understand evolutionary aspects of binary stars. Some reports from the 1920s even claimed the existence of a third body in the system. This could apparently fit some astrometric analyses of Sirius A and B, but has not been confirmed by other modern and very sensitive observations. I will also present in this part a compilation of most observational results gathered by modern astronomy.

The third part of the book (Chapters 6 and 7) will take the reader into the realm of space and non-optical astronomy, updating the level of knowledge as much as possible. It has been claimed that Sirius belongs to the family of the Sun, according to its proper motion. In particular, it was assigned membership in the Sirius-Ursa Major moving group, presumably a $\sim$300–500 Myr old stellar cluster. Recent publications refute this membership claim since the moving group appears significantly older than the Sirius system. α CMa is very close to the Sun, yet some intervening interstellar medium is detected. The interstellar space contains dense clouds of gas, grains of dust, hot plasma, etc.; these have been observed with the *Hubble Space Telescope*, with other space-based assets, and with ground-based astronomical facilities. It is interesting to consider how this could affect our observations of the Sirius stars.

The fourth and final part of the book (Chapters 8, 9, and 10) will consider the general question of stellar structure and evolution of stars like Sirius from a theoretical point of view. Single stars will be discussed first, then binary stars. I will present a compilation of the latest observational data about Sirius and of models proposed to explain this specific system, from which a coherent picture of the Sirius system and its history will, hopefully, emerge.

This is not intended to be a popular-level book. It assumes a certain amount of basic Astronomy and Physics knowledge, it contains quotations in Greek, Latin, and French (though I include English translations of these texts), and it does not explain very basic concepts. The readability and difficulty level of this monograph should not pose a problem for graduate students or for advanced amateur astronomers. Some parts, in particular those dealing with history and ethnography in the first part of the monograph, should be of interest to and be easily understood by lay persons.

Chapter 2
Historical perspective

2.1. Introduction

As night falls and the skies darken, the stars appear one by one until the dark vault fills with brilliant points of light. Among all stars, one stands out by its luminance. This is Sirius, $\Sigma\epsilon\iota\rho\iota\nu\sigma$ (Seirios) as it was called by the ancient Greeks. In order to find it one needs only to look South and to East of Orion, which is an outstanding autumn and winter constellation. In fact, the three bright stars of Orion's Belt make an excellent pointer to Sirius as Figure 2.1 shows: just project the line they define in an eastward direction and Sirius will be on the extension of this line.

Pointers are generally used in sky orientation to locate a specific object that might not be easy to identify; the stars of the Big Dipper, or the two outer ones of Cassiopeia, are used to locate Polaris, which is not so bright in its region to identify immediately. Note that using the Belt of Orion one can find the approximate location of Sirius even if it is below the horizon, since the distance to it approximately along this pointer direction is about seven times the length of the belt.

Very few Solar System bodies surpass Sirius in brilliance: the Sun, the Moon, and Venus are always brighter than Sirius, Mars and Jupiter are brighter than it in certain periods, but not all these bodies can be seen at all times. Sirius is, however, constant in its showing up above the horizon, if only one waits for the proper season and observes from the proper latitude (Sirius is not visible from the northern hemisphere above $\sim 74°$ North latitude, because its daily motion never takes it above the horizon). In fact, this star is so bright that it sometimes casts shadows (Parish 1985) and can be seen in daylight (Henshaw 1984). Its outstanding brilliance in comparison with its immediate neighborhood is demonstrated in Figure 2.2.

Because of its brilliance, Sirius has been worshipped in many civilizations. Most historical accounts emphasize its role in the Egyptian civilization; of this I will write more below. Robert Temple (1987) gave detailed accounts of folklore stories, which apparently link the cult of Sirius among some of the Mediterranean's ancient civilizations. Temple attempted to demonstrate that the various legends contain common beliefs, ranging from the Egyptian gods and the heliacal rising of Sirius to the twentieth century

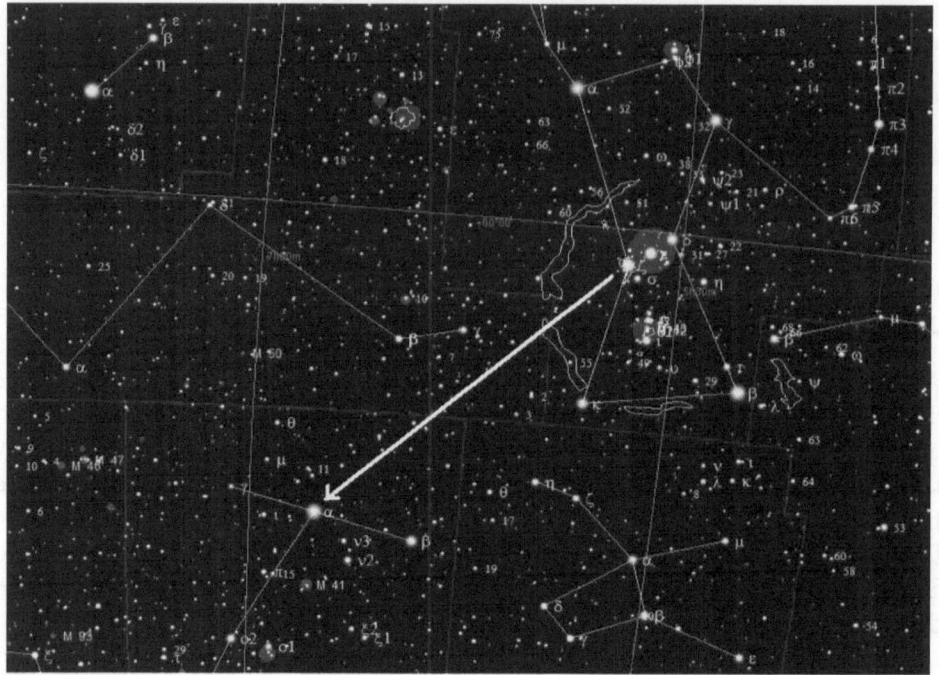

Figure 2.1. Pointing to Sirius, using Orion's belt. Sky chart produced with the *Cartes du Ciel* freeware.

beliefs of the Dogon tribe in the central African state of Mali. From this, he concluded that the evidence shows a case of extra-terrestrials visiting the Earth, specifically ancient Egypt. These extra-terrestrials he identified with the ancient gods who taught the people of Earth basic civilization skills; they are depicted by Temple as a race of amphibians. Since extraordinary claims require extraordinary evidence, these statements of Temple will not be discussed here any further, except for the consideration of the Dogon beliefs as reported by anthropologists (see below).

At this point, it would be fair to mention that Robert Temple answered many of the criticisms levelled at his *Sirius mysteries* book. This was primarily in a 1997 pamphlet available at his personal web site (http://www.robert-temple.com/) that contains an extended reply to Carl Sagan's criticism, as well as shorter pieces replying to other critics. I believe that none of his rebuttals is so compelling as to make one accept the proposal of extra-terrestrial cultural transfer, primarily in the light of the much easier accepted possibility of involuntary human cultural transfers during the nineteenth and twentieth centuries.

In most mythologies of the world, Sirius has always been connected either with some kind of divine being or, mostly by Northern Hemisphere civilizations, with a dog, a wolf, or a jackal (see Figures 2.3, 2.4 and 2.5).

Figure 2.2.   Sky photograph of the Canis Majoris constellation. The vertical extent
($\Delta\delta$) of the image is approximately $27°$ and the horizontal is about $2^h 25^m$ ($\sim 34°$).
Sirius is the brightest star that stands out considerably above all the other stars in this
image. Courtesy Akira Fujii/David Malin images. The image was obtained by Akira Fujii
on large format color-reversal (*transparency*) film using a high-quality standard camera
lens.

Note that Sirius is not the only bright star connected with a "dog
constellation"; not too far from the constellation in which Sirius resides,
the Big Dog (Canis Majoris, or CMa), one can find the Small (or Lesser)
Dog (Canis Minoris, or CMi). The brightest star of this constellation is
Procyon, "the one before the Dog", as the ancient Greeks called it. The
reason for this name is that Procyon, α CMi,[1] lies more to the North (and
to the East) than Sirius and thus rises before it; Procyon is therefore a kind
of harbinger of Sirius.

[1]Star identifiers can be proper names, such as "Sirius" or "Procyon", but by the
decision of the International Astronomical Union stars get official names. In the case
of very bright stars, the system devised in 1603 by Johann Bayer, and published in his
Uranometria, is retained. This system assigns single letters of the Greek alphabet to stars
of a constellation. The brightest star is called α, the second brightest is called β, and so
forth.

Figure 2.3.   Section of a map showing stars seen in the northern hemisphere drawn in 1469 by Giovanni Cinico. The Big Dog constellation is shown with a halo around the head to emphasize the brilliance of Sirius, and the dog runs after the darker lesser dog Procyon.

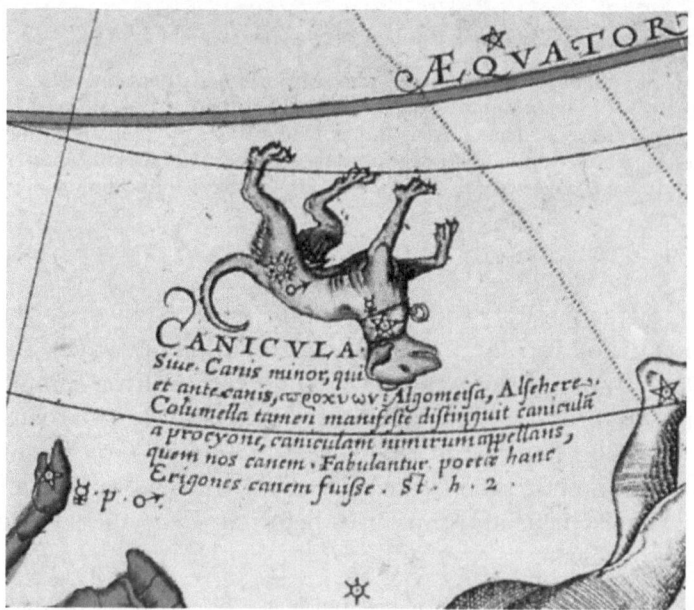

Figure 2.4.   Section of a map from 1590 showing the Canis Major constellation with the special mention of *Canicula*, i.e., Sirius.

2.2. Egypt and the ancient Middle East

Historical sources link Sirius intimately with the religion and day-to-day life of ancient Egyptians and, indeed, for calendaric purposes by many other ancient peoples. Well-defined risings or settings of Sirius could have served to make accurate calendars. Sirius is the brightest star in the sky, thus its appearance in the dawn sky just before the Sun (heliacal rising) would be a good starting point of the calendar. This is, of course, equally true for its setting just after the Sun (Kaler 1996), but this phenomenon of heliacal setting was not used for calendar keeping by ancient civilizations. Flammarion (1884) mentioned that the Egyptian calendar was regulated by the heliacal rising of Sirius; this event coincided with the summer solstice and with the annual flooding of the Nile (see below). One Egyptian name for Sirius, *Sothis*, means "brilliant" and Flammarion claimed that its predicting of the Nile flooding was represented as a warning dog.

Figure 2.5. Section of a map from 1709 that shows the Big Dog constellation with *Syrius* the bright star.

Flammarion (1884) pointed out a number of mysteries related to white Sirius, including that it was reported red in antiquity. Not only Sirius itself could have changed color, but other stars in the same Canis Majoris constellation became significantly brighter within ∼1,000 years. He mentioned the star 28 CMa as such; absent in ancient catalogs, it is known since the 1600s as a magnitude 4–5 object, clearly visible to the naked eye. However, Schaefer (private communication) found that in many cases stars of this magnitude range may be absent in ancient catalogs, such as

Ptolemy's Almagest. The general discussion on Sirius mysteries and, in particular, its possible color change, is deferred to a later section in the book (Chapter 3).

Sirius is intimately linked with life in ancient Egypt and, as already mentioned, it played a very important role in the definition of the Egyptian year. Its role in ancient Egypt was explained by Maffei (1981). The importance of this star cannot be underestimated, since it served to define the beginning of the year for this particular early civilization. Its role as calendar regulator was acquired because of the approximate coincidence, more than 5,000 years ago, of its heliacal rising as seen from Memphis and the beginning of the Nile flooding. This might have caused its possible inclusion in the ancient "Palette of Narmer", as Figures 2.6 and 2.7 show.

It is interesting that one needs to maintain astronomical records for a long time in order to detect temporal coincidences such as the heliacal rising of Sirius and the flooding of the Nile.

The rising of the Nile waters brought into the arable lands of the upper Egypt the fertile black mud, which allowed intensive cultivation of the land strip bordering the river Nile bringing wealth to the Egyptian state. The exact coincidence of the heliacal rising of Sirius along with the Nile flooding probably happened at ~4500 BCE[2] (Wells 1996). However, it is not clear at what phase of the month-long flooding was the peak, since the flood crest takes about 10 days to travel through Egypt (Holberg 2007). It is also reasonable to assume that seasonal changes, caused by Central African weather that modulate rainfall in the catchment basin on the Nile, would have influenced the yearly flooding.

The heliacal rising of a celestial body happens when this body appears above the Eastern horizon just before the rising of the Sun. This is the first time the star or celestial object can be seen after a period of a few months; before this special date the body is hidden in the glare of the Sun. As the Sun travels westward around the zodiac and completes a full circle every year, it changes its apparent position among the stars and allows some stars to become visible while others disappear in its glare. Sirius, for example, is not visible for a period of 70 days while the Sun's projected celestial position is close to it.

The need to predict the coming of the flood was essential to the development of the Egyptian civilization. The Sothic Cycle, and the synchronization of the calendar with the heliacal rising of Sirius, were therefore the key to both survival and successful civic management in ancient Egypt.

Since Sirius was important to the ancient Egyptians, its disappearance from the sky would have been disturbing. The 70 day period of Sirius invisibility, when the celestial location of the Sun is close to that of Sirius, played

[2]Dates are expressed here as BCE (before the common era) or CE (common era).

Figure 2.6.   The Palette of Narmer, dated at about 2925 BCE.

an important part in the process of mummification held in the highest regard by the Egyptians, the drying out of the corpse for its future preservation. The exact duration of this process was fixed at 70 days, even though it could have taken less time. Most Egyptian texts agree with Herodotus[3] that mummification took a total of 70 days from the death of the person until his burial. Thus, following the death of a prominent person, the ancient Egyptians set aside 70 days for the careful, ritualistic, treating and wrapping of the body of the deceased.

[3]Herodotus of Halicarnassus (∼484 BCE–∼425 BCE) was a Greek historian from Ionia.

Figure 2.7. Detail on the top part of the Palette of Narmer. The person following king Narmer is his sandal bearer, depicted with a star above his head; this star is sometimes interpreted as Sirius.

Sirius, called Sothis and Sopdet by the Egyptians, had various attributes such as the right eye of Ra (the Sun god *Ar-t-unemi*; Budger 1920), the crown on Ra's brow, the star which heralds the first day of the year, the Golden Sothis. The star was taken to be the heavenly representation of the goddess Isis, daughter of Ra and sister/wife of Osiris. A quote from Krupp (1983), with his comments in parentheses, related to the Hathor (or Het–Hert) temple at Dendera, describes the heliacal rising event:

> *Radiant rises the golden one (Hathor-Isis-Sirius) above the forehead of her father (near but in advance of the sun), and her mysterious form is at the head of his solar boat...As her fellow-divinities (the other stars) unite with her father's rays and as they merge with the glittering of his disk, Dendera is joyful...There is a festive mood as they behold the great one, the firmly striding creator of feasts in the holy city, on that beautiful day of the new year.*

Another aspect of Sirius was Isis-Hathor as the Great Mother-Creator. The heliacal rising re-established the world order by creating a new year, and Hathor, an ancient goddess, was described as:

...the beautiful one who appears in heaven, the truth which regulates the world at the head of the sun barge, the queen and mistress of awe, the ruler (of gods and) goddesses, Isis the great, the mother of gods.

Various images of Hathor from ancient Egypt have survived the ages. The depiction in the mortuary temple of Queen Hatshepsut, near the modern town of Der al Bahri, has Hathor-Isis represented as a standing cow with a disk (the Sun) between her horns (Allen 1963). Similar figures appear in the temple of Dendera and on the walls of the stepped pyramid of Sakkara, which dates back to ~2700 BCE (see Figures 2.8, 2.9 and 2.10).

The temple of Hathor at Denderah, situated 60-km north of Luxor on the west bank of the Nile opposite the provincial town of Qena, contains many inscriptions demonstrating the importance of Sirius for the ancient Egyptian civilization. The temple itself was probably oriented in a way to allow the viewing of Sirius. Lockyer (1894) found that no less than seven Egyptian temples were oriented in such a way so as to let the rays of the rising Sirius fall upon their altars.

Figure 2.8. The square zodiac from the Hathor temple at Denderah, now exhibited in the Louvre in Paris.

Figure 2.9.   Drawing of the square zodiac from the Hathor temple, from J. Bentley's
A Historical View of Hindu Astronomy (1823) Plate VIII, to help identify the figures in
the actual picture shown in Figure 2.8. Hathor is the reclining cow above the center and
to the right (*number 43*) with the star between its horns.

The importance of Sirius to the Egyptian civilization and mythology
is also demonstrated by the orientation of the star-shafts in the Great
Pyramid, built probably from 2589 to 2504 BCE. The shafts leading from
the Queen and King Chambers of the pyramid of Khufu pointed to specific stars: the southern shaft of the King Chamber to Alnitak (ζ Orionis)
with the Orion constellation being identified with Osiris, that of the Queen
Chamber to Sothis/Sirius identified with Isis. The northern shaft of the
King Chamber was probably oriented towards Thuban (α Draconis), the
Pole Star of the mid-3rd millennium BCE, and that of the Queen Chamber
to Kochab (β Ursae Minoris), the 'celestial adze of Horus'. Both northern shafts have a pronounced 'kink' imitating the shape of the opening-
the-mouth tool used in the final stage of entombing the mummy. The
opening-the-mouth, fertility and other rites were probably performed in
these 'rebirth chambers' of the pyramid to impregnate Isis=Sirius, to enable
the soul of the deceased king to depart to Orion, and to help the birth of
Horus (Malek 1994).

Figure 2.10. Detail from the square zodiac from the Hathor temple at Denderah, showing Sirius as a star between the horns of the cow Hathor.

The ruins of the Denderah temple were discovered by Napoleon's officers in 1798. The foundation of the temple was dated to ∼1790 BCE (Brown 1971) although it was rebuilt on the same site from the Middle Kingdom (2040–1640 BCE), with the last modification done by the time of the Roman emperor Trajan. The existing structure was erected no later than the late Ptolemaic period, around the first century BCE. Its peculiar square zodiac (Figure 2.8), documented by members of Napoleon's expeditionary army in Egypt, was "removed" in 1820 by Sebastien Saulnierin and moved to France with the permission of Egyptian ruler Mohamed Ali Pasha. The square zodiac is now displayed in the Museum of the Louvre. The zodiac depicts astronomical events (solar and lunar eclipses) that took place around 50 BCE. It shows α CMa in the form of a five-pointed star between the horns of a cow carried in a boat (Figure 2.10). The boat is followed by the goddess Satet, carrying a bow and an arrow pointed at Sirius, an image probably linked with the Nile flooding. We shall see later that other civilizations adopted a link between Sirius and a bow-and-arrow configuration. Despite the very significant role played by Sirius in regulating the Egyptian calendar, it is notable that the *Catalogue of the Universe*, compiled in Egypt around 1100 BCE, did not list this star (Thurston 1994).

Chapman-Rietschi (1995) also mentioned that the Nile flooding and the heliacal rising of *Sopdet* or *Sepdet*[4] (Sirius) coincided with the beginning of

---

[4]The ancient Egyptians wrote in Hieroglyphs, graphical icons that do not include vowels. Thus the exact written form for Sirius is transliterated as *spdt* and one can only guess as to how was this name actually pronounced.

the Egyptian New Year. However, he found no link between Sopdet and any kind of dog as one would expect given that it belongs to the constellation Big Dog; he quotes from Kákosy's *Lexicon der Aegyptologie* that Sopdet apparently means "sharp" or "pointed". The Egyptians called the heliacal rising of Sirius *Peret Sepdet*, the "going forth" of Sirius. This reminds one of the apparent similarity between the Egyptian *P.R.T.* and the Hebrew root *prtz* (peretz[5]); this has a similar meaning of "erupting".

The ancient Egyptian calendar was based on lunar cycles, as are the Jewish and Islamic calendars at present. The month was defined in relation to the New Moon, starting when the waning crescent disappeared from the pre-dawn sky (Krupp 1983). The lunar cycle has approximately 29.5 days between New Moons, thus the Egyptians had months of 29 and months of 30 days. A purely lunar calendar of 12 cycles of New Moons is 11 days shorter than the solar year; without correction the annual events would occur later every year.

The Egyptians corrected the calendar by adding an extra (intercalatory) month whenever required and this was decided according to the heliacal rising of Sirius. In the Middle Kingdom (2030 to 1640 BCE) the last month of the year was called *Wep-renpet*, which means "Opener of the Year". If the heliacal rising happened during the last 11 days of *Wep-renpet*, an additional month called *Djehuty (Greek name: Thoth)*[6] would be added at the beginning of the new year. This was done to avoid having the festival of Wep–renpet (and the rising of Sirius) from falling into the first lunar month of the next year. This thirteenth month would need to be added approximately once every third year.

The Egyptian calendar was apparently in place as early as 3100 BCE, in the pre-dynastic period. The evidence for this is an ivory plaque dated from the First Dynasty, with carvings of a cow symbolizing Sirius and the Nile flood. This indicates that the coincidence of flooding and heliacal rising was noticed even before the establishing of a institutionalized monarchy in the United Egypt,[7] and emphasizes the necessity of accurate seasonal prediction for a primarily agrarian society. The Egyptian calendar was reformed around 1200 BCE to the Sothic form of 12 months per year, each month being 30 days long, with five additional festive days. Note that Thurston (1994) puts this calendar revision further back to ∼3000 BCE.

The revised Egyptian year had three "stages" each of four months length (Tetramenia), and five additional "festive" days. The year started always on 1 Thot, the first day of the first month of the Egyptian year. This was

---

[5]In Hebrew the vowels are normally not written but are implied from prior knowledge.

[6]The name of the Egyptian god connected with record-keeping, writing, and the Moon.

[7]The unification of the kingdoms of Upper Egypt and Lower Egypt took place at the time of the famous King Narmer at approximately 3100 BCE.

a fixed calendar, which corresponded approximately to the tropical year.[8] The stages of the year, linked to the main activities of a primarily agrarian society, were *Akhet* (inundation or flooding of the Nile), *Peret* (emergence or growth of crops) and *Shemu* or shomu (summer, Nile low water and harvest, a very hot and dry period in the year). Present-day farmers in Egypt, the *fellahin*, still use this three-season division of the year.

The almost six hours difference between the Egyptian year and the tropical year caused the date of the heliacal rising of Sirius to slip by one day every four years. Even more important is the slippage during a long period of time; it takes $365.25 \times 4 = 1461$ "fixed" Egyptian years for the date of the heliacal rising to circle completely through the year! In order to keep the calendar synchronized with the seasons, i.e., with the solar year, the heliacal rising of Sirius was used. This period, when 1 Thot coincides again with the Nile flooding and with the heliacal rising of Sirius, marks therefore the "Sothis period". The Roman historian Censorinus wrote in the third century CE that the heliacal rising fell on Egyptian New Year in 139 CE. If the heliacal rising of Sirius and the beginning of the year really corresponded at the time the Egyptian calendar was introduced, it implies that the years 1322 BCE, 2782 BCE, or even 4242 BCE were beginnings of Sothis cycles.

The dates of the heliacal rising of Sirius were calculated by Schaefer (2000). The exact dates of heliacal rising of Sirius, as viewed from a site at latitude 30° North, were calculated by Schaefer as July 16 for the year 3500 BCE, July 17 in 1500 BCE, and July 20 in 500 CE. Now this event takes place on August 4. At the beginning of the 40 day period Sirius sets just after sunset and it is very close to the Sun in the sky. Then it disappears in the Sun's glare for some days, and at the end of the period starts rising just a bit before the Sun does. Schaefer mentioned that the actual date of a specific heliacal rising could vary by a few days, depending on the actual atmospheric extinction and on other random factors.

Not only in Egypt was Sirius important. In the ancient civilized world outside Egypt Sirius was mostly identified with celestial canine creatures. A minor Babylonian god was named "Sirrush" (Ceram 1967). Sirrush is depicted as a dragon-like creature, sacred to Marduk and accompanying him (at least in some pictures). The sirrush is the creature depicted on the reconstructed Ishtar Gate of the ancient city of Babylon and is shown in Figures 2.11 and 2.12. It resembles a scaly dragon with hind legs like an eagle's talons and feline forelegs. It also has a long neck and tail, a horned head, a snakelike tongue and a crest. One wonders whether there might be a link between Sirius and Sirrush, although some Greek etymologies made from Sirius "Seirios" ($\sigma\eta\iota\rho\iota\sigma\sigma$).

---

[8]The tropical year is the time interval between consecutive passages of the Sun through the spring equinox, the First Point of Aries, and has exactly 365.242447 days.

Figure 2.11. Sirrush, the "dragon of Babylon", as depicted on the reconstructed Ishtar gate of Babylon exhibited at the Pergamon Museum in Berlin (Image downloaded from the Wikipedia, free for use under the Creative Commons Attribution ShareAlike 2.5).

Figure 2.12. Sirrush of the Ishtar gate from the Pergamon Museum in Berlin shown on an East German stamp.

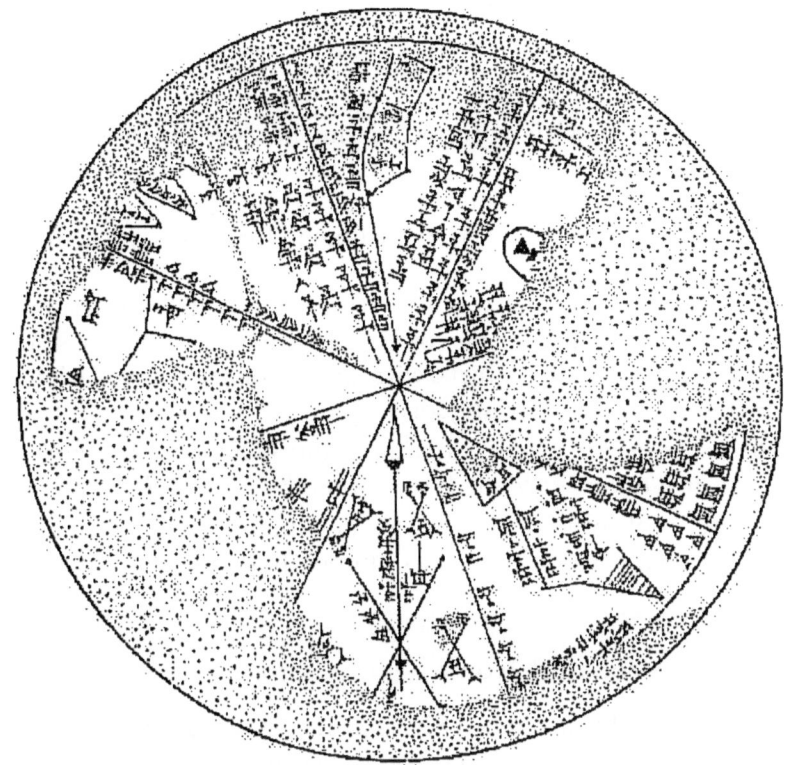

Figure 2.13.   Ancient Nineveh star map shows Sirius as the sharply pointed arrow.

Sirius itself was called "the Arrow" both by ancient Summerians and ancient Babylonians (Britton & Walker 1996; see Figure 2.13). The arrow was apparently fired from Orion's bow. The later Persians referred to Sirius as *Tir*, also meaning "the Arrow." The Phoenicians called it *HaNabeah* (the Barker), implying a canine connection. According to Burnham's Celestial Handbook (Burnham 1966), Sirius was known to the Assyrians as *Kal-bu Sa-mas* (the Dog of the Sun). Its old Akkadian name was *Mul-lik-ud* (Dog Star of the Sun) and in Babylonia, *Kakkab-lik-ku* ("Star of the Dog"). In Chaldea, the star was it Kak-shisha ("the Dog Star that Leads") or *Du-shisha* ("the Director").

Later, in the same part of the world, Sirius was taken to be a dog also in the Mithraic imagery, as reported by Chapman-Rietschi (1997). He mentioned an opinion that Mithraism took Sirius to represent a blood-thirsty dog as the controller of heavens.

Another ancient Near East religion, Zoroastrianism, stated in its sacred the Zend-Avesta book that the Supreme God established as Master and Overseer of all the stars the star Sirius, called *Tishtriya*.

I will sacrifice unto Tishtrya, the bright and glorious star, whom Ahura Mazda has established as a lord and overseer above all stars,... (Khorda Avesta III 44)

An association of Sirius with water occurs in the Zend-Avesta, which calls Tishtriya (Sirius) the responsible for rain and the enemy of the daemon of dryness. According to the Zend-Avesta, the Supreme God once made Tishtriya cause a massive flood as punishment for men's wickedness and corruption.

In ancient Babylon, Sirius was sometimes lumped together with the planets. A document from the time of king Esarhaddon (681–668 BCE), a treaty with a Medianite king, starts with an astrological invocation of the planet-gods and of the territorial gods:

In the presence of the planets Jupiter, Venus, Saturn, Mercury, Mars, Sirius, and in the presence of Assur, Anu, Enlil, Ea, Sin, Shamash,...

Sirius, therefore, took precedence, in this document, even before *Shamash* (the Sun) and *Sin* (the Moon), and was lumped together with Solar System objects.

2.3. Ancient Greece and Rome

Greek and Roman traditions linked Sirius primarily with canine personages, and with the weather. Serviss (1908) mentioned in his book an early-Greek ivory disk, found by Schliemann at Troy, with a representation of a dog believed to be the constellation CMa. Schliemann's book (Schliemann 1880) contains indeed a report and a reproduction of a small ivory disk on p. 585, but this disk does not show an image of a dog.

The Iliad of Homerus (book XXII), where the Trojan war is described, has Achilles likened to Sirius:

Him the old man Priam first beheld, as he sped across the plain, blazing as the star that cometh forth at harvest time, and plain seen his rays shine forth amid the host of stars in the darkness of the night, the star whose name men call Orion's Dog.

The "blazing" is probably the shiny bronze armor Achilles wore, likened to the splendor of Sirius not only in the brilliance but also in the flickering of its light and its breaking up into a multi-colored display, and seen on Sirius depending on the steadiness of the atmosphere (Serviss 1908).

Much about the sources of the name Sirius can be found in Allen (1963). He quoted Eratosthenes[9] as saying that:

[9]Erathostenes of Cyrene (276–194 BCE), Greek mathematician, astronomer and geographer active in the famous school of Alexandria that started to operate in Egypt during

Such stars astronomers call σειρίους (seirious) on account of the tremulos motion of their light.

From this, we learn that the Greek term Σείρ (Seir) and its derivations were probably connected with great brilliance in general, and with the high brightness of Sirius in particular. However, Serviss (1908) linked this Greek name to the Egyptian God named *Osiris* and even mentioned the possibility that the name is related to the Celtic word *Syr*.[10] This Celtic connection is also mentioned by Allen (1963), but note that the Scots called Sirius *Reul-an-iuchair*.

Sirius is so brilliant that, in ancient times, it warranted special mention on sculptures. The Atlas Farnese, a marble sculpture of the giant Atlas holding the world on his shoulder (Figure. 2.14), depicts the Universe as a celestial globe with *bas-reliefs* of the important constellations. Canis Majoris is represented as a dog, and the location of Sirius among the stars of the constellation is marked by rays of light emanating from the head of the Big Dog. The sculpture was claimed to be the ∼200 CE copy of a globe constructed by Eudoxus of Cnydos or Cnidus[11] and is exhibited in the Museo Nazionale Archeologico in Naples (Brown 1971).

Parenthetically, note that in antiquity marble sculptures used to be painted before being exhibited. A close look at color reproductions of the Atlas Farnese does indeed show faint traces of paint on the globe, but it is difficult to see from a reproduction what was colored and in which color. The globe depicts 42 of the 48 classical Greek constellation figures but not the stars comprising each constellation. The stars may have originally been painted on the globe. The observations, on which the constellations depicted on the globe are based, were made probably in 125±55 BCE, ruling out the possibility that the Farnese Atlas would be a copy of a globe made by Eudoxus. Schaefer (2005) attributed the observations to Hipparchus. To date, no study of this celestial globe with an intention to check for color traces of the different objects depicted there has been undertaken. This has some implications on the question of the redness of Sirius, which will be discussed below.

In the Astronomica of Hyginus[12] Sirius is described as part of the Dog constellation.

the rule of Alexander the Great, at about 322 BCE. He was one of the first to calculate the size of the Earth.

[10] Seviss does not provide references to this in his book, but attributes the possibility to a person named Dupuis. I could not find a Celtic dictionary to check this up, but in modern Welsh "Syr" is simply "Sir" (Jon Davies, private communication).

[11] Eudoxus of Cnidus (410 or 408 BCE to 355 or 347 BCE) was a Greek astronomer, mathematician, physician, scholar and student of Plato.

[12] Gaius Julius Hyginus (ca. 64 BCE to 17 CE) was a Latin author.

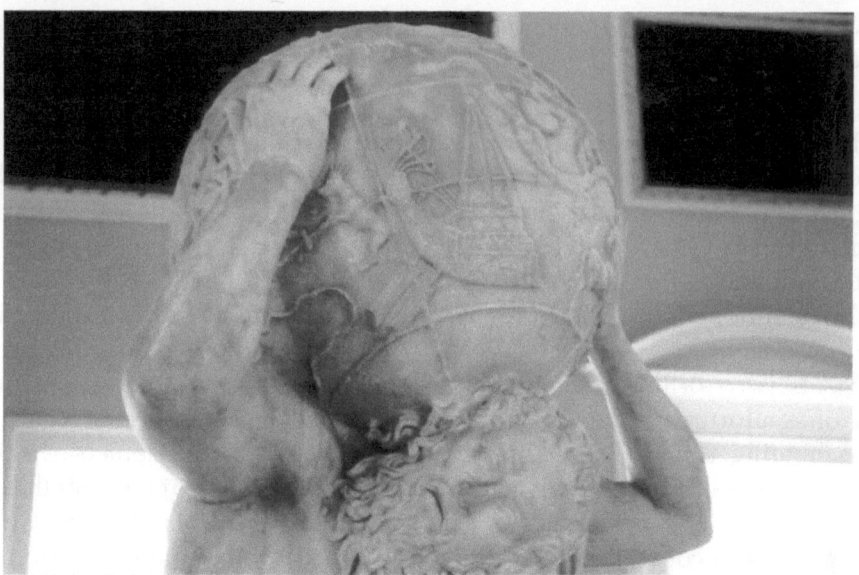

Figure 2.14.   Enlargement of the celestial globe on the shoulder of the Atlas Farnese sculpture shows Sirius as the bright star at the head of the Big Dog, with rays of light emanating from it (just off the palm of Atlas, by the prow of the ship). (Photographed in 1998 by Dr. Gerald Picus and used with his permission).

The Dog has one star on his tongue which itself is called Dog, and on its head another which Isis is thought to have put there under her own name, and to have called it Sirius on account of the brilliance of the flame because it seems to shine more than the rest. So, in order for men to recognize it more easily, she called it Sirius.

One of the earliest sky maps on which Sirius and Canis Majoris are drawn is the Geruvius planisphere, dated from the second century CE and shown in Figure 2.15. This sky map contains mythical figures described in the *Phaenomena*[13] of Aratus. Although not astronomically correct, since it shows CMa just below Cancer, it depicts a large dog with the name of the star inscribed below the figure: *Syrius*. Curiously, between it and the constellation of Orion (recognizable because of its name and because of the rabbit "Lepus" following it) there is another figure of a dog with the word "Anncanis" over it. Assuming that *Anncanis* could mean ante-canis, before-the-dog, this would make the smaller dog *Procyon* that indeed rose before Sirius the Dog for Alexandria of the second century CE, even though

[13] A poem composed around 280 BCE by the poet Aratus of Soli. Apparently a long versification of an astronomical work by Eudoxus dated 370 BCE (Brown 1971), this poem includes a description of 43 constellations.

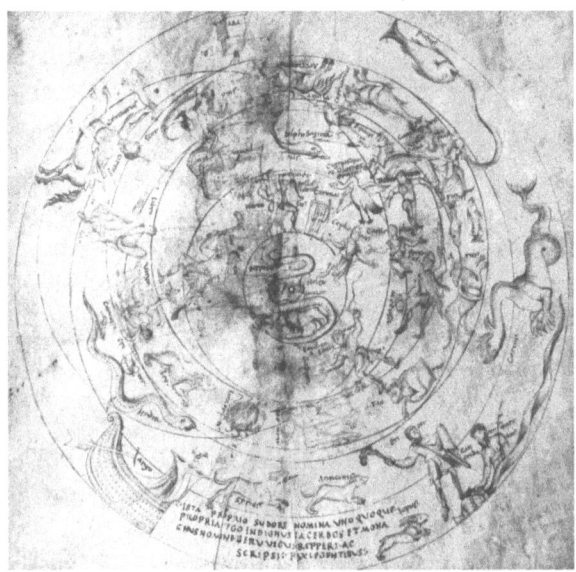

Figure 2.15. The planisphere of Geruvius, from ~2nd century CE, shows two dogs near the bottom end and above the writing. The leftmost larger dog is labelled *Syrius*. It chases a smaller dog called *Anncanis*.

it is further East of it. In any case, it seems that Geruvius took Sirius to be a constellation by itself.

An Aratus reference links Sirius to heat:

> *A star that keenest of all blazes with a searing flame and him men call Seirios. When he rises with Helios (the Sun), no longer do the trees deceive him by the feeble freshness of their leaves. For easily with his keen glance he pierces their ranks, and to some he gives strength but of others he blights the bark utterly. Of him too at his setting are we aware.*
>
> (Aratus, Phaenomena 328)

The heliacal rising of Sirius was taking place in antiquity during the hottest season of the year, when only mad dogs would venture out in the open. This is why the Romans use to call the period near the heliacal rising *dies caniculariae* or Dog Days. In principle, this period of about 40 days long extended from 20 days before to 20 days after the conjunction of Sirius with the Sun, from about July 3 to August 11, and was shorter than the 70 days of Sirius invisibility for the Egyptians since Rome is further North than Egypt. At the end of this period Sirius rises just a bit before the Sun; this latter event is the heliacal rising described above.

Sirius is mentioned a number of times in Hesiod's[14] *Works and Days*
long poem dealing with agricultural works, where Sirius acts on the daily
life. In the summer Sirius changes both women (to become libidinous) and
men (to become weaker) by the addition of its rays to those from the Sun:

> *But when the artichoke flowers [in June], and the chirping grass-hopper
> sits in a tree and pours down his shrill song continually from under his
> wings in the season of wearisome heat, then goats are plumpest and wine
> sweetest; women are most wanton, but men are feeblest, because Sirius
> parches head and knees and the skin is dry through heat.*

However, all this is past and things calm down in the autumn with the
gathering of the grapes:

> *But when Orion and Sirius are come into mid-heaven, and rosy-fingered
> Dawn sees Arcturus [in September], then cut off all the grape-clusters,
> Perses, and bring them home.*

Later in the year Sirius also signals the best time for wood harvesting:

> *When the piercing power and sultry heat of the sun abate, and almighty
> Zeus sends the autumn rains [in October], and men's flesh comes to feel
> far easier, – for then the star Sirius passes over the heads of men, who
> are born to misery, only a little while by day and takes greater share
> of night, – then, when it showers its leaves to the ground and stops
> sprouting, the wood you cut with your axe is least liable to worm.*

Hesiod was using celestial phenomena as calendar markers, as identified
by Aveni (2002). These are conveniently presented in Aveni's Table 2.1,
where Sirius is mentioned not only during its heliacal rising but also when
passing on the meridian while Arcturus rises heliacaly in mid-September.

Hippocrates, the Father of Medicine, made much of the influence of
Sirius. He was a Greek physician who lived from approximately 460 to 377
BCE. He included a discussion of Sirius in his *Corpus Hippocraticum* in *de
Hebdomadibus*, describing the power of this star over the weather and the
consequent physical effect upon mankind. The rising of Sirius bisected the
summer and the more fatal fevers happened between the risings of Sirius
and Arcturus than elsewhere in the year (West 1971).

The Romans, like the ancient Greeks, regarded Sirius highly when dealing with medical issues. Galen (130–~200 CE), who studied medicine and
practiced in many places including Rome (attending the emperors Marcus
Aurelius, Commodus, and Severus), accepted the proposition that stars in
general influence human affairs. Galen, in his book *On Medical Experience*,

---

[14]Hesiod or Hesiodos was an early Greek poet who presumably lived around 700 BCE,
in the time of Homerus.

considered the position of Sirius in the sky whenever medicine was being prepared or administered.

Pliny was a Roman scholar of science, politics, and literature who lived from 23 to 79 CE. He also accepted the view that Sirius was highly influential upon our planet. Oddly enough, Pliny also associated Sirius at the times of its rising with the origin of honey:

> *This substance is engendered from the air, mostly at the rising of the constellations, and more especially when Sirius is shining...*
> (The Natural History Book XI 12)

And also with its quality and sweetness!

> *If the honey is taken at the rising of the Sirius, and if the ascent of Venus, Jupiter or Mercury should happen to fall on the same day, as often is the case, the sweetness of the substance and the virtue which it possesses of restoring men to life, are not inferior to those attributed to the nectar of the gods.*   (The Natural History Book XI 14)

Seneca,[15] in his Oedipus tragedy of the first century, linked Sirius to extreme heat as did Greek authors before him:

> *No soft breeze with its cool breath relieves our breasts that pant with heat, no gentle Zephyrus blows; but Titan [Helios the sun] augments the scorching dog-star's [Seirios'] fires, close-pressing upon the Nemean Lion's [i.e. Leo, the zodiac sign of mid-summer] back. Water has fled the streams, and from the herbage verdure.*   (Seneca, Oedipus 37)

Diodorus Siculus[16] wrote in his *Bibliotheca historica*, in the description of Aristaeus:

> *To this island [Ceos] he sailed, but since a plague prevailed throughout Greece the sacrifice he offered there was on behalf of all the Greeks. And since the sacrifice was made at the time of the rising of the star Sirius, which is the period when the etesian winds customarily blow, the pestilential diseases, we are told, came to an end.*

Ceragioli (1996) mentioned an interesting ritual performed in the third and the second centuries BCE by the inhabitants of Ceos in the Aegean: men of the island would dress up in armor and ascend a hill to witness the heliacal rising of Sirius. Priests would sacrifice to Zeus and Sirius, praying to receive the northern cooling wind. The aspect of Sirius at rising would serve as an omen; if the star would be dim and hazy, the year would bring sickness. If Sirius would appear brilliantly clear at rising, the year would

---

[15]Lucius Annaeus Seneca (∼4 BCE to 65 CE) was a Roman Stoic philosopher, statesman, and dramatist.
[16]Greek historian, ∼90 BCE–∼30 BCE.

be healthful. This emphasizes the link the Greeks made between Sirius and diseases.

The auspicious influence of the star Sirius on human beings was carried into the field of healing even in the late Middle Ages. The astrologers of that time felt that the heliacal rising of Sirius was an outstandingly favorable time for the gathering of medicinal herbs, while the alchemical physicians of the late 1500s recommended that certain ingredients be gathered or captured throughout the year at the rising of Sirius. Sirius was thought to rule savine, mugwort, dragonwort, and the tongue of a snake, all serving as ingredients in medieval magic or medicinal compounds. Indeed, it appears that the medieval view of Sirian influence included certain curative or healing powers, as those qualities were definitely attributed to the influence of Canis Major. In the late 1650s CE the rising of the star Sirius was thought to coincide with the agitation of snakes, an association probably connecting serpents with places of healing.

The Greek and Roman traditions of connecting Sirius and its constellation with dogs were preserved later in the names of constellations, as shown by the *Alfonsine Tables*[17] that call Sirius *Canis Syrius*. Other connections of Sirius with a dog, mentioned by Serviss (1908) are in the Scandinavian mythology *the dog of Sigurd*, in India *the Deerslayer*, and in ancient Greece as *the hound of Acteon* and *one of Diana's hunting dogs*.

Flammarion (1884) linked the Greek word *seirios* with the Sanskrit "svar", meaning "shining=[French]briller" and "lighting=[French]éclairer". He connected the early-morning rising of Sirius, a few thousands of years ago, near mid-June and which heralded the Nile flooding in ancient Egypt, with the "Dog Days". These days begin, according to Théon of Alexandria,[18] 20 days before the heliacal rising of Sirius and end 20 days after this special rising. During this period it was believed that dogs get mad and humans get fevers; this is why Roman authors like Virgilius, Horatius, and Manilius recommended that people get out of the cities during these days. By the way, similar beliefs of leaving cities during difficult times were customary also much later: for example, the personages in the Decameron of Boccacio left their town for the country because of "pestilence", and so did the young Newton leave Cambridge at the behest of his parents returning to Woolsthorpe, his family's village, during the plague epidemic of 1665.

Flammarion (1884) remarked that the ancient link between the Dog Days and the hot season is no longer valid, because the heliacal rising of Sirius takes now place around the end of August, whereas the Dog Days,

---

[17]Astronomical tables drawn up at Toledo by order of Alfonso X around 1252–1270.
[18]Teacher of mathematics and astronomy in Alexandria (Egypt) ∼335–405 CE. Reported on the solar eclipses of June 16, 364 and November 25, 364.

at least as far as nineteenth century almanacs go, were still listed as the period between July 3 and August 11 every year.

2.4. Africa and Arabia

Even in the southernmost parts of Africa Sirius was an important celestial object. Warner (1996) wrote that one possible name for Sirius in the Zulu language is *Intsanta*, which means a star which is "specially bright and scintillates numerous rays". The Basuto people, on the other hand, put Sirius in the constellation *Magakgala* together with Rigel, Betelgeuse, and Procyon, not emphasizing it by itself.

Snedegar (1997) wrote that the Zulu believe that:

Indosa is a star which rises before morning star, when night is advanced; and if men have stayed drinking beer, or eating the meat at a wedding feast, if they see Indosa arisen, for it arises red, they say, 'Let us lie down; it is now night'

Thus a red Sirius rising would indicate that the night is not yet over, but its white rising would announce the imminent morning.

Another special custom linked with Sirius was mentioned by Warner (1996) and concerns the San and the Khoikhoi tribes, hunter-gatherer people of the South-Western Africa. They performed special ceremonies when Sirius first appeared that required the rapid movement of burning sticks pointed at Sirius. Warner assumed that this apparently imitated the twinkling of the star when near the horizon. However, I believe that this could also symbolize the kindling of embers; when these get in contact with more oxygen they become whiter and brighter. These customs demonstrate the special attention given to Sirius even in the very distant parts of Southern Africa.

Sirius, along with six other stars (β Triangulus=Tri, η and α Tau, γ, ι, and κ Ori), was claimed to be used to synchronize the calendar in northwest Kenya, near Lake Turkana, by a Cushite people at about 300 BCE. The two relevant sites, Namoratunga (or Ng'amoritung'a) I and II, are about 210-km apart and consist of alignments of massive standing stones. The Namoratunga II site has 20 peculiar polygonal basalt columns. The identification of the alignments with the position of relatively bright stars at rising was done by Paul (1979). He showed that only during a relatively narrow period of time, from 430 BCE to 150 BCE, could all seven alignments operate simultaneously. This period coincides with the carbon dating of the site, at 2,285±165 years before present.

In early Arabia Sirius was called *Al Shi'ra al 'Abur al Yamaniyyah*, the Brightly Shining Star of Passage of Yemen, probably because it set in the direction of that province. Arabian astronomers called it *Al Kalb al Akbar*, the Greater Dog, following the Greek and Latin traditions, and Al Biruni[19] named it *Al Kalb al Jabbar*, the Dog of the Giant, directly from the Greek perception of the figure linking the Big Dog with Orion the hunter.

Much later, in the Holy Qur'an, Sirius is mentioned by name, the only star for which this is done: "That He is the Lord of Sirius (the Mighty Star)" in Surat An-Najm (The Star: Sura 53), verse 49, though this is sometimes translated as "Lord of the galaxies". In general, however, the Arabs followed the Greco–Roman tradition described below of associating Sirius with a dog and called it *Al Kalb al Akbar* (the great dog).

The Bedouins of the Sinai and the Negev, however, called Sirius *Al-Burbarah*, probably deriving the name from the barking noise of a dog (Bailey 1974). They used its appearance on the easterly horizon at nightfall, in mid-January, as a season marker for the start of their 40-day winter season *ash-shita* that lasts till late-February.

Bailey (1974) quoted from the Bedouin tradition:

Al-Burbārah, limmā taṣīr li-Suhayl riṣhi
'Iṣhī walā hū íṣhī
Awwal rabi' wi ākhar iṣhtī

['when Sirius hangs over Canopus like a bucket rope
At the very beginning of the evening
'Tis the last of winter and the first of spring].

The calendaric use of Sirius by the Bedouins of Southern Sinai is reported by Bailey (1974) as connected with the ripening of dates:

Ath-Thurayyā, razn -ignī
Imagaydiḥ, bahī
Al-Jawzā, zahī
Al-Burbārah, jinī

['Under the Pleiades, the branch is heavy
Under Aldebaran, (the date) shines
Under Betelgeuse, (the date) darkens
Under Sirius, (the date) is ripe'].

The Bedouins, therefore, used Sirius and other bright stars as the simplest markers of their practical calendar. Bailey (1974) writes that in the Arabian desert, this astronomical knowledge did not "spring from indigenous roots" but incorporated a few local legends in knowledge imported mainly from Greece and India. The use of stellar markers allowed the regulation of annual agricultural activities despite the use of the Muslim lunar calendar whose months rotate among the seasons.

---

[19]Abu Rayhan Muhammad ibn Ahmad al-Biruni (973–1048) was a Persian Muslim universal genius. He wrote no less than 35 books on astronomy, and many on other subjects.

2.5. India, China, and the Far East

There is no evidence that Sirius served any calendaristic role in Eastern civilizations, but it appears that, as an exceptional star, it attracted significant attention. Here also canine references abound: Chapman-Rietschi (1995) mentioned an ancient Chinese mortuary vessel in the Metropolitan Museum of Art in New York, dated from the early Western Han dynasty (∼205 BCE), which is decorated with a blue beast with bared fangs lunging at a mounted archer. The blue beast (wolf) represents Sirius, known in China as the Heavenly Wolf, and the archer represents the adjoining constellation Bow, with an arrow always pointing directly to the Wolf star.

The place of Sirius among the Chinese asterisms was explained by Staal (1984). Its name was *T'ien Lang*, the Celestial Jackal. Interestingly, some of the other stars in CMa formed, for the ancient Chinese, a hunting dog *T'ien-Kaou*, the celestial dog who was hunting the jackal.

Staal (1984) wrote that when the Chinese moved South, and more stars became visible over the southern horizon, a new asterism was invented. This was *Hou-Chi*, the bow and arrow, where the short arrow points (almost) straight at Sirius. Note the similarity of this Chinese constellation with that of the ancient Babylonians; in both cases this was a bow-and-arrow combination. The Babylonian one was called "bow and arrow" (^{mul}BAN, Chapman-Rietschi 1995), and it contained the star δ CMa. Sirius was at the tip of the arrow (Brecher 1979).

At this point, it is worth noting that the Chinese system of constellations was completely different from the Western one. As early as the third century CE the Chinese had 283 constellations composed out of 1,464 stars (Stott 1995). However, the canine connection for Sirius was apparently retained with the revision of the constellations.

The Hindus held Sirius to be a Rain God, and the Chaldeans called it "Star of the Dog". The sacred Hindu texts of the Vedas, written probably in ∼1500 BCE, refer to Sirius as *Tishriya*; this is similar to the Persian *Tishtrya* discussed previously.

2.6. North and South America

Among North American cultures Sirius served as one of the pointer stars, presumably for calendar and seasonal keeping. At least some of the medicine wheels, for instance, the Bighorn Medicine Wheel (Figure 2.16) on top of the Bighorn Range in Wyoming, contain a sight-line claimed to be pointing to the rising location of Sirius (Eddy 1974). Other sight-lines featured by medicine wheels point to Aldebaran (α Taurus=Tau), Rigel (β Orionis=Ori), Fomalhaut (α Pisces Austrinus=PsA), and to the Sun's position on special equinoctal or solstitial dates. This explanation, of medicine

Figure 2.16.   Reconstruction of the circle of stones with sight-lines, after the original medicine wheel at Bighorn in Wyoming, erected at the Valley City State University in North Dakota. One sight-line of this ancient observatory, probably used by Native Americans to predict the seasons, was claimed to be pointing to the rising location of Sirius.

wheels being astronomical calendars, has been disputed by a number of investigators based on astronomical, statistical, and anthropological grounds.

Schaefer (2000) disputed the alignment of the Big Horn medicine wheel based on accurate calculations of the visibility of Sirius. As he explained, when Sirius is exactly on the horizon it can never be seen as a bright star because of the atmospheric extinction. In order to be perceived as a bright object Sirius must be at least $\sim 5°$ above the horizon. In this case, its azimuth will also be $\sim 5°$ south of what most investigators assume, thus the spokes of the Bighorn medicine wheel could not be pointing at its perceived heliacal rising point.

One possibility, however, is that the observers would have accepted the heliacal rising even if Sirius would have been seen as a faint star given the high extinction, provided that faint star would be located at the expected position (on the projection of Orion's belt, see Figure 2.1).

Holberg (2007) mentioned that the Pawnee called Sirius "the Wolf Star" or "the White Star", one of the four pillars that hold the sky up. In the Pawnee mythology Sirius would accompany the dead souls along the Milky Way, similar to the Inuit belief. Other Pawnees knew Sirius as the Coyote star.

The Inuit people in Alaska knew Sirius as "the Moon Dog" and believed that its conjunction with the Moon brings strong winds. To the Cherokee it is one of the dog (or wolf) stars that guards the Milky Way together with

Antares on the other side of the Milky Way. *Agise'gwa*, the Great Female Sirius, and *Wa'hyaya'*, the Alpha Male (Antares), must be fed or they will not allow a soul to pass. It will get caught between the two, and will wander endlessly back and forth across the Great River in the Sky.

In South America, the Desana people of Colombia have a complex set of beliefs concerning stars and the structure of the society (Aveni 1996). In this context, Sirius served as an absolute boundary separating unmarried from married life. In a cosmic conspiracy theory, one could ask whether this belief was somehow linked with Egyptian beliefs about Sirius as being at the boundary between years.

2.7. Polynesia and Australia

It is probably only a coincidence that the name of the Sun in Polynesia is *Ra*, as in Ancient Egypt, while Sirius itself is called *A'a* by the Polynesians, despite Thor Heyerdahl's claim of cultural links between the two cultures that could cause one to expect a name close to *Sepdet*.

Holberg (2007) noted that, since the declination of Sirius is approximately $-17°$, it passes overhead of Fiji. This star could, therefore, be used for navigation purposes as well as a season mark. Its appearance in the morning sky, which marked the Roman *Dies Caniculares* of extreme heat, symbolized for the New Zealand's Maori people the freezing cold of winter.

The Australian aborigines, as did people elsewhere in the world, knew Sirius as a constellation by itself, calling it *Eagle* (Brown 1971). For the Koori Mara people who live in the states of Victoria, Tasmania and the southern parts of New South Wales, the star Sirius was the female wedge-tail eagle, *Gneeanggar*, who was carried off by *Waa* the crow (Canopus).

2.8. Jewish connections

Allen (1963) wrote that the Hebrews knew Sirius as *Sichor*, its Egyptian name. He also attributed the Ideler[20] (no reference given) the claim that the adoration of the Se'irim, which is specifically prohibited to the Jews in Leviticus 17:7, may have referred to the cult of Sirius and Procyon. The word *Se'irim* is usually translated as "devils" by e.g., the King James version of the Bible: *And they shall no more offer their sacrifices unto devils, after whom they have gone a whoring*, or as "goat demons" by other popular English translations of the Old Testament: *And they shall no longer sacrifice their sacrifices to the goat demons with which they play the harlot.*

---

[20]Presumably Christian Ludwig Ideler (1766–1846), German chronologist and astronomer.

The connection seems to be more with the Greek word σειρίους (seirious) than with an Egyptian word root.

Ehrlich[21] (1959) mentioned that the name of Sirius in Hebrew is *Avrek* (the Shining). He added that, in the 1930s, astronomers found it to be not a double system, but a triple one, with a third star orbiting the fainter companion with a period of one-and-a-half years. This claim will be discussed below.

I could not find an explanation for the name *Avrek*, except that the word may be mentioned with a slightly different spelling in the Bible (Genesis 41:43) where when Joseph is honored by Pharaoh by getting the Pharaoh's second-best chariot, the people are told to bend their knee:

And they cried before him "avrech".

This might be an Egyptian loan word and its meaning would then be uncertain. It could mean "look out" in Egyptian, or possibly "to your knees" from the Hebrew *berech*=knee. This last explanation seems to be supported by the Talmud, but does not appear to be connected with Sirius at all. On the other hand, a connection with the Hebrew word *barak* is more likely, since this means "shiny", and would imply a word invention by Ehrlich (1959).

2.9. Conclusions

Sirius was a star of interest for many civilizations. The above attributions are probably only the tip of a much larger cultural and historical legacy concerning Sirius. It may be that this was because of its exceeding brilliance, or because the star served a practical purpose, such as predictor of the Nile flood in Egypt. In many civilizations, from the Far East to the Western Europe through ancient Greece and Rome, the star was identified with a dog, a jackal, or a wolf. This was not the case in ancient Egypt. In any case, the attention the star received ensured its continuous observation throughout the ages. A comparison of its observed properties, as recorded by different histories, may be helpful in understanding some evolutionary or other observational features of Sirius in particular, and of stars in general.

Note also the relatively high public interest in Sirius in recent years, partly derived from the mystical properties that this star is believed to possess and partly as a result of its perceived high brilliance taken to imply a specific importance for human activity. This can be emphasized by quotations from astrological literature and web sites. The two-word combination "Sirius" and "astrology" brings up more than 790,000 results on Google; "Sirius" and "astronomy" brings up more than one and a half million.

---

[21] Asher Ehrlich was a teacher, amateur astronomer, and popularizer of astronomy in Jewish community living in Palestine, in the first half of the twentieth century.

However, while the mystical interest in Sirius is obvious, it is important also to note that Sirius served as a source of inspiration for artists. J.K. Rowling elected to give her secondary Harry Potter character, wizard and godfather of Harry Potter, Sirius Black, the name of the star and a last name which begins with the letter designating the companion star Sirius B (the binary character of Sirius will be discussed below). Sometimes the wizard Sirius Black even turns into a black dog.....

Sirius was also the subject of a musical piece for tuba and piano, composed by Allen Sap and performed at the University of Cincinnati College-Conservatory of Music on February 3, 1985. A number of poems have this star as a subject.

Chapter 3
Mysteries of the Sirius system

This chapter discusses three issues considered as specific mysteries connected with Sirius: its red color reported by some ancient sources and some possible explanations for it, hints to its binary nature derived from old observation, and the amazing knowledge about the Sirius system of a tribe in French West Africa reported by some investigators.

3.1. The issue of historical redness

Some tantalizing historical documents argue that perhaps Sirius, now an obviously white star, was red in antiquity. A very comprehensive discussion of historical evidence for the redness of Sirius is contained in a series of papers by Ceragioli, in particular his review of Greek and Roman literature on Sirius published in the Journal for the History of Astronomy (Ceragioli 1995). In a nutshell, his thesis is that (a) the literary evidence is clear that Sirius was perceived as red by Ptolemy and Seneca, (b) many of the other historical evidence items presented on this subject are more wishful thinking than real evidence, and (c) it is not clear that the literary evidence should be taken to imply physical evidence. In other words, Ceragioli thinks that Sirius may have been red only sometimes, at rising and setting for example, and that would suffice for it to be perceived as red. He also connected Sirius to the Dog Days of the summer, explaining thus a connection with heat, and implying that this connection attached the red color mention to Sirius, the bringer of heat.

I shall describe Ceragioli's historical evidence in the chronological order in which the various articles have been published. Much of the material is in his highly recommended 1995 paper and is not repeated here, but the more important issues are, along with his interpretation of this evidence. At the end of the historical presentation I shall address briefly some explanations put forward to account for a possible changing color of Sirius.

The first person to mention a red Sirius, whose color changed in historical times, was Barker (1760) when discussing the "mutations of the stars" (see also Lynn 1887a and See 1927). In 1839 Herschel mentioned Sirius as historically red (according to Malin & Murdin 1984 and to Ceragioli

1995). This was because Herschel became convinced that stars could change after observing the outstanding eruption of η Carinae in the mid-1800s. An extensive discussion on the red Sirius issue can be found also in Flammarion (1884), where he explained the Greek word "seirios" as implying "shining" and "burning", as already mentioned above.

The question of the color of Sirius was raised by Flammarion in the context of Ptolemy's stellar catalog, part of Ptolemy's compendium Megale Syntaxis (Μεγάλη Σύνταξις), dated from 138 CE and known also as the Almagest[1] (e.g., Heiberg 1898). Fomenko et al. (1993) quoted the name of this book as μαθηματικη συνταξις (mathematike syntaxis) and asserted that the last observation included there had been made on February 2, 141 CE. Ptolemy is Claudius Ptolemaus, the scientist and philosopher who worked in Alexandria (Egypt) from 130 to 175 CE (or 127–141 to 168 CE; Fomenko et al. 1993). The name of his compendium was passed on by the Arabs as *Al Kitab al Magisti*, becoming *Almagestum* in Latin and *Almagest* in English.

Ptolemy's Almagest is a 13 volume work that attempted to concentrate all the astronomy, cartography and mathematics known until Ptolemy's time in a single publication. The stellar catalog was apparently a compilation of all astronomical observations available to Ptolemy, with possibly some observations made by Ptolemy himself, and makes up most of the *7th* and *8th* volumes of the Almagest. The Almagest also describes how a stellar globe should be built. The observations reported in the Almagest date as far back as the time of the Babylonian king *Nebu-na-sir* (747–734 BCE), presumably near the beginning of Babylonian astronomy (Britton & Walker 1996).

Ptolemy wrotes of Sirius (quoted from Flammarion):

ο έν τω στόματι λαμπρότατος κυων καί ύπόκιρρος

The intention here was definitely to describe Sirius, because the quoted passage refers to the mouth (στόματι) and to the Dog (κυων). The last word, transliterated here as "ippokiros", is intended to mean "reddish" in Greek, although Flammarion presented also an alternate explanation, which he assigned to M. Schjellerup, that the two last words of Ptolemy's description should have been καί σείριως, transforming the sentence from:

The brilliant, on the mouth, called the Dog and reddish

to the more banal:

The brilliant, on the mouth, called the Dog and Sirius

---

[1]This is an Arabization of the Greek name, enhancing in part the importance of Ptolemy's composition. This is because of the use of the Greek word *megistos* (Μεγιστος), which means "Greatest", instead of the original, which means simply "Great".

However, the literal interpretation of a reddish color is the one mostly accepted.

The interpretation of the term used by Ptolemy to describe the color of Sirius was explained by Ceragioli (1995). He wrote that the Greek term used in the Almagest, ὑπόκιρρος, derives from the simple Greek adjective κιρρός (kirros). This adjective was used, according to Ceragioli, to describe shades of wine, and is presumably the equivalent of the "blush" or rosé wines known today. Specifically, Ceragioli mentioned a progression of wine colors from white, through yellowish, to pale pink (which is *kirros*), to red wine, and finally to a dark red or purplish shade. The prefix ὑπο should imply, therefore, "somewhat" or "faintly", which can be adopted here as the colloquial "kind of". Along with the term *kirros*, it probably implied a pale shade of pink, perhaps not really deep red.

Ptolemy referred to Sirius also in book VIII of the Almagest, when he explained how to mark a celestial globe, starting with Sirius. This advice was "because it [Sirius] is the brightest star in the sky" (Evans 1987). Ptolemy suggested making the globe dark in order to show off the colors of the yellow stars, and presumably the red of Sirius, the first star to be marked on a celestial globe (Ceragioli 1996). The Farnese globe, perched on the shoulder of Atlas and discussed in the previous chapter, seems to carry a trace of color on some of the *bas-relief* features, as already mentioned; perhaps some of the stars had been painted on the newly produced celestial globe that now shows only the constellations, in the same manner as other antique statues were painted.

A problem with Ptolemy's color information is that the mention of a reddish Sirius appears only in the Almagest. The other treatise Ptolemy compiled and where stellar colors are discussed, the Tetrabiblos treatise of astrology, mentions only Antares (α Sco), Betelgeuse (α Ori), Arcturus (α Boo), Aldebaran (α Tau) and Pollux (β Gem) as red, but not Sirius. Therefore, Ptolemy himself was not consistent regarding the question of the true color of Sirius at the time the Almagest was composed.

Flammarion (1884) presented also the following evidence for ancient redness from Seneca, a philosopher born in Spain at the end of the first century BCE and the tutor of the Roman emperor Nero:

> *Nec mirum est, si terrae omnis generis et varia evaporatio est, quam in coelo quoque non unus appareat color rerum, sed acritor sit Caniculae rubor, Martis remissor, Jovis nullus.*

which translates as:

> The variety of the emissions of the terrestrial atmosphere should not surprise us. Even in the sky the celestial bodies exhibit different colors: the stars of the Canicula shine with a lively red, Mars is paler, and Jupiter has no specific color.

Canicula here means, apparently, the star Sirius itself and the quote indicates that it was redder than Mars. The brilliant star Sirius was sometimes taken to be a constellation by itself (see below). Brecher (1979) asserted that this statement by Seneca can be taken as unambiguous; Seneca apparently was a *sharp observer who seems to have interpreted astronomical observations more correctly than Aristotle.*

The 1887 issues of the British journal *Observatory* saw some intense correspondence regarding the color of Sirius and the alleged color change in historical times. This apparently started when Lynn (1887b) drew attention to the translation by Schjellerup of the description of the fixed stars by Al Sufi,[2] probably referred to also by Flammarion (1884). Lynn's argument, by which he claimed that Al Sufi did not see Sirius as red, was that the treatise does not refer to Sirius as ὑπόκιρρος. Moreover, Lynn added that Al-Battani[3] mentioned that Ptolemy spoke only of five red stars, while the surviving editions of Ptolemy's Sytaxis mentioned six stars carrying the red indicative: Arcturus (α Boo), Aldebaran (α Tau), Pollux (α Gem), Antares (α Sco), Betelgeuse (α Ori), and Sirius (α CMa). This is the difference, already mentioned, between the Megale Syntaxis and the Tetrabiblos. Also, Lynn took exception with the translation of the Greek text and mentioned the same possible erroneous translation as did Flammarion.

Five years later, new discussions appeared in the *Observatory* regarding historical color changes. Lynn (1902) presented the evidence from Hyginus,[4] who called Sirius *propter flammae candorem*, which could mean either white or bright. He then followed with a quote from Plautus[5] about Arcturus, where this star was called "splendens stella candida". As Arcturus is patently a red star, Lynn argued that the adjective "candida" should be interpreted as "bright", not as "white", when ascribed to stars. Both Gare (1902) and Oom (1902) agreed that Seneca's statement mentioned by Flammarion (1884) is categorical and imply a red Sirius.

The continuation of discussions on the color of Sirius was taken up by the US astronomer T.J.J. See.[6] Ceragioli (1995) mentioned that the professional career of See was ruined soon after it began: in 1892 he was appointed instructor at the University of Chicago after completing a PhD in Berlin with the mention *magna cum laudae*, but was fired from this

[2] Abd el-Rachman bin Umar al Sufi, Persian astronomer (903–986), who wrote the Book of Fixed Stars in the tenth century CE and dedicated it to his friend and pupil, the ruler of Persia and Iraq, Sultan Adud el-Daulah.

[3] Abu Allah Mohammad ibn Jabi ibn Sinan al-Raqqi al-Harrani al-Sabi, ∼855–929 CE. Arab astronomer who lived and studied in Syria. Known also as Albategnius, he composed a work in astronomy with a more accurate description of the motion of the Sun and the Moon than given by Ptolemy. His star catalog, part of his *Kitab al-Zij*, lists 489 stars.

[4] C. Julius Hyginus, Roman astronomer active in the first century BCE.

[5] Titus Maccius Plautus, ∼254–184 BCE, was a Roman comic poet.

[6] Thomas Jefferson Jackson See (1866–1962).

instructorship in 1896 after quarrelling with G.E. Hale, the director. He became an assistant at the Lowell Observatory, and was fired from there in 1898 after antagonizing the staff. His professorship at the US Naval Observatory ended in 1902 after a nervous breakdown with a "banishment" to the naval time-keeping station at Mare Island, California. The reason for passing through so many jobs was primarily a fully-blown self-esteem that caused conflicts with many of his peers and with his superiors. One could derive from Ceragioli's history of See that this was a person whose statements could not be fully trusted. However, one should also note that in addition to being an excellent student in his early days, See's work on binary star orbits was very much appreciated by his peers.

See (1927) summarized the historical evidence from published documents, leading to a claim that Sirius changed color between the epochs of Ptolemy (second century CE) and al-Sufi (tenth century CE). He expressed dissatisfaction with Schjellerup's translation of al-Sufi in which the critical number of five red (bright) stars was mentioned. He found that both Albategnius (al Battani, mentioned above) and Alfarganus[7] were silent when Ptolemy's observations of red stars were mentioned, and the same was true for Al Sufi and Ulugh Beigh (or Begh).[8] These, being the foremost Arab and Persian astronomical authorities of their times, caused See to conclude that there is no reason to doubt that Sirius was red in antiquity and that it changed color subsequently.

See continued to Ptolemy, exalting the good observing conditions from Egypt, where he visited in 1891, and the presumably excellent conditions from the terrace of the Alexandria Museum where Ptolemy supposedly observed the sky from 100 to 140 CE. See observed both Sirius and Canopus (α PsA) from Egypt and remarked on their similar colors; his conclusion was that Ptolemy could not have been wrong about the red color of Sirius because of some atmospheric effect. In Ptolemy's time, Sirius would have culminated in Egypt more than $42°$ above the horizon; much higher than Canopus, or α and β Centauri, all which are not called red stars by Ptolemy. Therefore, most of the time Sirius would have been visible far from the "reddening" conditions near the horizon.

See presented an extensive comparison of Greek and Roman sources in his 1927 paper, which supports the idea of a red Sirius. His table (on p. 270

[7] Abu'l-Abbas Ahmad ibn Muhammad ibn Kathir al-Farghani (c. 860-c. 920 CE), was one of the astronomers and engineers in the service of Khalif al-Mamun. al-Farghani wrote the book *Kitab fi al-Harakat al-Samawiye wa Jawami Ilm al-Mujun* (The Book on Celestial Motions and through Science of the Stars), translated into Latin in the twelfth century.

[8] Muhammed Taragai Ulugh Begh (1394–1449) was the grandson of Timur the conqueror and ruled over Transoxonia, the region between the rivers Amu Darya and Syr Daria, in today's Uzbekistan. He was one of the most important observational astronomers of the fifteenth century and the builder of the famous observatory in Samarkand (now Uzbekistan).

TABLE 3.1. Historical references supporting a red color for Sirius (in Mediterranean civilizations, after See 1927)

Source	Period	Reference to Sirius	Meaning
Euripides	~440 BCE	πυρός φλογές (pouros floges)	flame colored
Aratus	~270 BCE	ποικίλος (poikilos)	colored
Cicero	~50 BCE	Rutilo cum lumine	with ruddy light
Virgillus	~30 BCE	Ardebat in coelo	burning in the sky
Horatius	~10 BCE	Rubra Canicula	red dog
Germanicus Caesar	~19 CE	urgetur cursu rutili Canis	fast races a reddish dog
Manilius	~20 CE	Rabit suo igne	raging to join its fire
Seneca	~25 CE	Acrior sit, Caniculae rubor, Martis remissior, Jovis nullus	forceful is the redness of Sirius, less than that of Mars, Jupiter has none
Pliny	~70 CE	Ardore Sideris, Sirio ardente	burning of the sky, Sirius burning
Columela	~100 CE	Sirius ardor	burning of Sirius
Geminus	~100 CE	πύρινος (pouros)	flame colored
Ptolemy	~150 CE	ὑπόκιρρος	reddish
Avienus	~400 CE	multus rubor imbuit ora	much redness that fills the region
Théon	~450 CE	ποικίλος (poikilos)	colored

of the paper) is partially reproduced here as Table 3.1, with dates for the various sources added and with comparisons between Sirius and Antares deleted.

Note that, in reference to Germanicus Caesar,[9] See mentioned that the work he referred to is the translation of the book written by Aratus, probably made by Germanicus when he was governor of Syria. As Germanicus was a military person, See assumed that he was well-versed in observing the sky, thus his words should be given suitable attention.

See (1927) added to his discussion the mention that ruddy dogs were sacrificed both in Rome and in Greece, to ward off the pestilence associated with the Dog Star. The evidence of the color of sacrificial dogs, he concluded, points to a red Sirius at least till the beginning of the Christian era. The period of redness could have ended as early as the fourth century, if the evidence from Théon and Avienus[10] is to be trusted; this could be a major change in the star's appearance that took place within only 300 years.

[9]Roman general and consul, 15 BCE to 19 CE. Was the grand-nephew of the emperor Augustus and the adopted son of the emperor Tiberius.
[10]Postumius Rufius Festus Avienus (or Avienius), fourth century CE, translator of the Phaenomena of Aratus of Soli into Latin.

At about the same time as See's 1927 paper was published, another piece of evidence of antique sightings of a red Sirius was presented by Stenzel (1928). In a discussion of ancient Egyptian references on Sirius he mentioned that its name (Soped, or Sothis) seems very common on Egyptian monuments as one could expect, given its calendaric and religious significance. The hieroglyph depicting Sirius contains a triangle; this appears in the inscriptions with a red surface. Since the colors on Egyptian monuments were not selected at random but were selected according to their symbolic significance, Stenzel considered this a strengthening argument for an ancient red Sirius. He found another pointer to a red Sirius in antiquity in that Isis was depicted in the Dendera temple as a goddess adorned with a red veil.

A dissenting view about a red Sirius in antiquity was presented by Plassmann (1927). In a reference to the *AVESTA*, the holy book of the Persians, he quoted from the eighth sacrificial song:

We honor the magnificent Tischtrya-star *(Sirius) the white shines with his bright unparalleled rays.*

As *AVESTA* dates probably from the fifth century BCE, Plassman concluded that at that time Sirius was not red. No sooner was this Eastern refuting reference published, that another one, pointing in the opposite direction, appeared. Dittrich (1928) found evidence in Babylonian writings for a red Sirius. He quoted the writing on a stele dated from king *Asur-nasir-apal* (885–860 BCE) that these were the days of *KAK.SI.DI*, which shone as red as copper. *KAK.SI.DI* was apparently the name Babylonians gave to Sirius; it formed the tip of the arrow in the depiction of one of their constellations as Figure 2.13 show (Brecher 1979).

The entire question of antique observations and references to a red Sirius was revised anew by Pannekoek (1961). He translated Ptolemy's $\upsilon\pi o\kappa\upsilon\rho\rho o\varsigma$ as "yellowish" and linked this to the summer season and to the Roman *rubra canicula*, but concluded that the color was due only to atmospheric extinction.

Another mid-Eastern reference for the possible redness of Sirius is in the Cerberus Slab of Hatra (Tuman 1983). The slab is a *bas-relief* of painted limestone, discovered in the early 1900s in the temple of Hatra[11] and exhibited in the Mossul Museum, in Iraq. The slab is dated to the second half of the first century CE and is a product of the Parthian civilization.[12]

Tuman (1983) identified astronomically-valid designs on the slab. In particular, he claimed that the three dogs held in leash by a giant man-like figure shown in Figure 3.1 are the three stars Sirius (α CMa), Adhara

---

[11] A ruined city located in the Al-Jazira region of present-day Iraq, about 110 km South–West of Mossul and 290 km North–West of Baghdad. Hatra was designated a UNESCO World Heritage site in 1985.

[12] The Parthian empire ruled from 247 BCE to 228 CE in ancient Persia. At one time, this empire ruled areas extending from present-day Turkey to Pakistan.

Figure 3.1.   Segment of the slab of Hatra. According to Tuman (1983), the giant figure represents Orion the hunter, and the middle dog of the three, colored reddish on the slab, symbolizes Sirius.

(ϵ CMa), and Procyon (α CMi). The giant was identified by Tuman as Orion the hunter. The colors of the various items in the *bas-relief* are not discussed in Tuman's paper, yet the three dogs are depicted with different colors: black, white, and red (Tuman 1998, private communication). If there is any meaning to be attached to the color, one could be that the red dog was the one symbolizing Sirius in the first century CE. However, note that while ϵ CMa is definitely a white star, Procyon is far from being a "black" star. Its color is rather yellowish, since it has a similar spectral type to that of the Sun.

The 1980s focused the attention of the astronomical community on China and the East. Swings (1982), reporting on Chinese astronomy, mentioned a quote from Sima Qian, a "Historical Record" from the first century BCE, quoted by Fang Li-Zhi as "the white is like Sirius". Van Gent (1984) noted that from the first century BCE to the seventh century CE the ancient Chinese texts called Sirius a white star.

Schlosser & Bergmann (1985) found additional information on the color of Sirius in a Lombardian manuscript dated to the eighth century CE, authored by Gregory of Tours[13] called "De cursu stellarum ratio" and preserved in the Bamberg library. The manuscript was probably produced in order to instruct the monks in the performance of night religious services, and the timing of these services was reckoned by astronomical observations. Sirius is identified in the manuscript by the times of rising and setting. It is called *stella splendida*, a unique characterization in this manuscript, and its proper name is given as "Rubeola" or "Robeola", meaning "rusty".

Van Gent (1986) took exception to the claim of Schlosser & Bergmann (1985). He asserted that the times of rising and setting given by Gregory of Tours could not have fit Sirius, because the object must have been visible throughout the year, from 1 h every night in September to eight during the winter months. This visibility pattern could fit Arcturus (α Boo), a northern red star, a much better candidate than Sirius which lies in the southern celestial hemisphere and traces only a short arc above the horizon for observers in Tours. In particular, Van Gent mentioned that the star is supposed to have been visible twice during the night in September: once shortly after sunset and again just before sunrise. This, he wrote, would again fit better Arcturus, but not Sirius.

The same argument, that *Rubeola* is Arcturus, was also suggested by McCluskey (1987). He also proposed to identify the constellation Gregory calls *Quinio*, as Sirius and five stars in its vicinity. Van Gent (1987) repeated his assertion that Robeola is Arcturus, this time basing it on the hours of visibility. Schlosser & Bergmann (1987) replied to both McCluskey and Van Gent rejecting their identification of Robeola with Arcturus on the basis on an unequal duration of the day and night hours during the period of bishop Gregorius and on the recognition that Gregorius was, apparently, an amateur astronomer not particularly interested in the accuracy of numerical data.

Ridpath (1988) quoted from Manilius's[14] *Astronomia*, written in ∼15 CE:

Vix sole minor, nisi quod procul haerens frigida caeruleo consorquet lumina vultu

which means that Sirius is hardly inferior to the Sun, "save that its abode is far away and the beams it launches from its sea-blue face are cold". Ridpath took this to mean that Sirius was not red in the first century CE, but blue.

---

[13]Georgius Florentinus Gregorius (∼538–∼594 CE), bishop of Tours, where the council of Tours took place in 567 CE. The treatise was apparently written between 575 and 582 CE and was intended to present practical ways to comply with the decree issued by the council.

[14]Marcus Manilius, Roman didactic poet in the first centuries CE.

However, Bicknell (1989) translated the Latin of Manilius as implying the blue of the sky, where somewhere else Manilius wrote:

magna fides hoc posse color cursus que micantis ignis ad os

meaning that there is strong proof that the constellation has this power is the color and the path of the fire gleaming at its mouth. This presumably returns us again to the star in the mouth of the Dog, i.e., Sirius. Since the star brings war, Bicknell argued that its color must have been red.

Ceragioli (1995) brought forward a quote from Hephaestion of Thebes,[15] originally found by Schiaparelli and published in his 1897 rebuttal of See's arguments. Hephaestion reported on Egyptian observations of Sirius that they were deriving omens from the star's color at heliacal rising. This is a report of practices as true for about 2 BCE (Van Gent 1984). If Sirius would rise

> *great and white and if it came through with its colour, then the Nile's current would rise and there would be abundance; but if it was flame-coloured [πυρρός] and the colour of red ochre [μιλτώδης] there would be war*

Schiaparelli concluded, rightfully, that in order for Sirius to show off all these colors at heliacal rising its intrinsic color should have been white.

Gry & Bonnet-Bidaud (1990) published a quotation from the second century BCE, from the history book *Shiji* Chapter 27 dealing with the Han dynasty. The book was written by the historian and astronomer Sima Qian (145–87 BCE). The quote refers to the color of Sirius and the original is reproduced here as Figure 3.2. It is rather difficult to obtain a uniform translation, thus I present here three versions. Gry & Bonnet-Bidaud translated the passage as follows:

At East there is big star called Wolf/ Wolf horn changes color/,/many thieves robbers

The quote and translation are repeated in Bonnet-Bidaud & Gry (1991), where they added the information that Sima Qian "collected all historical and scientific material available at his time after the burning of books ordered by the emperor Qin Shihuangdi in 213 BCE". Bonnet-Bidaud & Gry remarked also that the term "horn" could suggest some visible asymmetry in the aspect of Sirius at that time.

It is possible that the passage was mis-translated. I asked Dr. Marcella Contini, an astrophysicist at Tel Aviv University and reader of Chinese, for an independent translation of the writing. She produced the following somewhat cryptic paragraph:

> *It is established by imperial order by four books stored in the (Zen) storehouse that at the East exists a star. The brilliance of Sirius has*

---

[15]Greek-speaking Egyptian astrologer in the fourth and fifth century CE.

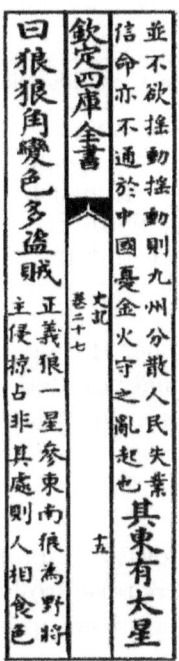

Figure 3.2.   Chinese ideograms of the poem describing the colors of Sirius.

many colors - strong right principles of astrology that it is in the East are defeated by a good appearance in the South. It is supported on the contrary that the color is on point to lead.

A further translation of the same passage, reproduced by Ceragioly (1995) from Johnson (1961, quoted as 1962 by Ceragioli), is:

When the points of Lang change color, there will be more robberies and thieves...When the color is yellow or white and bright, it signifies good fortune; if the points turn red, military disturbance will arise

The text refers apparently to Sirius, called *Lang* (Wolf) in Chinese and none of the three versions seems to indicate any specific color that could be associated with the it. The quote supports a view that the ancient Chinese did not attach a particularly dominant color to Sirius, but rather observed its fast changes of colors.

It is possible to question the consistency of Chinese sources that refer to the colors of stars not only in the context of Sirius, but also regarding other objects. For instance, the entry of the "Historical Record" by Sima Qian, from the first century BCE (Swings 1982), which is often quoted as a reference for a historically white Sirius, is listed in Table 3.2 together with modern catalog colors of the stars it mentions.

TABLE 3.2.   Chinese historical reference for star properties, from the first century BCE

Ancient Chinese	Name & classification	Catalog B–V	Present color
The white is like Sirius	α CMa: A0m	0.00	White
The red is like Antares	α Sco: M1.5Iab-Ib	+1.83	Red
The yellow is like Betelgeuse	α Ori: M1-2Ia-Iab	+1.85	Red
The blue is like Bellatrix	γ Ori: B2III	−0.22	White
The dim is like Mirach	β And: M0$^+$IIIa	+1.58	Red

One wonders, therefore, whether this "description" should not be interpreted as indicating a color change for Betelgeuse (α Ori), from yellow to patently red in 2,000 years! Also, why should one consider Mirach dim, when it is now much brighter than the faint limit of visual sighting? Its present-day visual magnitude is +1.75, while the faintest visible star is below +6.0.

Not only was the above quotation about the colors of stars as perceived by the Chinese astronomers taken to be an argument against the red color of Sirius, but also, and in a mixed-up way, another quote from Sima Qian could have been interpreted as indicating the color white. Chapman-Rietschi (1995) quoted a translation by Chavannes in 1898, which compares the color of Sirius to that of the planet Venus:

When Tai bai (Venus) is white it resembles the star Lang (Sirius)

This could mean that Sirius was white, or it could mean that it was bright and that, when Venus was showing as white, it was as bright as Sirius. Therefore, this specific quote could not be construed to be a strong argument in favor of a white Sirius.

The ancient Hawaiian navigators used bright stars to find their way at sea, as mentioned in the previous chapter. One particular asterism they employed was *Ke Ka o Makali'i* (The Canoe Bailer of Makali'i), made up of the stars *Hoku-lei* ("Star-Wreath"=Capella [α Auriga]), *Na Mahoe* ("The Twins"=Castor [α Gemini] and Pollux [β Gemini]), *Puana* ("Blossom"= Procyon [α Canis Minoris]), and *'A'a* ("Burning brightly"=Sirius). Another Hawaiian name for 'A'a was apparently *Kaulu-lena* or *Kaulua-lena*, meaning "Yellow star". It is possible that the Hawaiian navigators perceived Sirius to be yellow, a color different from but close to the red reported in and around the Mediterranean basin. Yet they clearly saw that Aldebaran [α Taurus] was a red star, as its Hawaiian name indicates: *Hoku'ula* meaning "Red star", and also Betelgeuse, called by them *Kaulua-koko* ("Brilliant red star").

Sirius is not the only star in the sky which some historic sources qualify with a color different from what it has these days. Condos & Reaves (1971)

mentioned that Altair (α Aquila) was red sometime before Ptolemy compiled the Almagest. Hind[16] observed a star (HD 38451) changing color in 1851; Warner & Sneden (1988) looked at Hind's variable object and found that it is a shell star[17]. The Chinese document *Shi Chi*, dated at ~100 BCE, recorded a white or yellow color for Betelgeuse (α Ori), as reported by Swings (1982) and Tang (1986), and as explained above. However, the color of Betelgeuse is now red, and so it has been reported by Ptolemy in the Almagest and in the Tetrabiblos. Flammarion (1884) mentions that Al Sufi saw Algol (α Per) red in the tenth century; it is now white and so it was in the time of Ptolemy.

Is it at all possible that one could see the color of Sirius? Is this star not too bright for that? The entire question of color perception of stars, including the physiology of vision, which is relevant when considering historical evidence for a color change in Sirius, was discussed by Steffey (1982). He claimed that only ~150 stars show colors to the unaided eye. The brightest of these, some 30 in number, show up as yellowish-orange, yellow, or pale blue. Only the very red stars look indeed red; these are usually carbon stars[18], are faint, and require some visual aid (binoculars, for example) to notice their colors.

Given the sensitivity of the eye to different colors, Steffey (1982) mentioned that blue stars are difficult objects in which to discern color already from $2nd$ magnitude and fainter. In order to be able to see yellow, albeit with a somewhat brownish tinge, one can observe stars as faint as $3rd$ mag. Very bright stars will always look white; Sirius, which is a bluish star today since it has the same B–V color as Vega, appears white to the eye because of the saturation of the color receptors in the eye (the cones in the retina). In order to see Sirius as a red star, one therefore requires it to be fainter than it is today by at least 1.5 mag (but no more than ~4 mag to still be considered a bright star) and redder by at least one magnitude.

Schaefer (private communication) argued that it would be possible to see the color of Sirius despite the statement by Steffey (1982), since the light from the star would be spread over the retina to the resolution of the human eye, some 1 arcmin across. In this case, the surface brightness of the image would be sufficiently low so as not to saturate the color receptors in the retina. In fact, the average surface brightness would be some 4 mag fainter than that of the full Moon; since one can obviously see colors on the lunar surface, it follows that it is also possible to see the color of Sirius.

[16] John Russell Hind was an English astronomer in the nineteenth century.

[17] A shell star is an object in a transitory stage of evolution, after it has ejected a small fraction of its envelope. The ejected matter produces observable signatures by absorbing some of the starlight.

[18] Carbon stars are stars that show absorption bands of carbon molecules in their spectra. Most are red giants and exhibit stellar winds. Their surface compositions have changed during the post-main-sequence evolution.

3.2. Explanations for redness

If one accepts that Sirius was perceived as red in antiquity and is white today, then only an astronomical explanation makes sense. This is because other causes, such as a significant change in the response of the human vision to color during a few thousands of years, seems very unlikely. Such astronomical explanations have been proposed by all who reported on a historical red color. Except for actual physical changes of Sirius B from a red giant to a white dwarf, which was mentioned above, or a nuclear fusion process taking place on the surface of the white dwarf companion, these involve extinction either in the Earth's atmosphere or in the interstellar space between us and Sirius. I shall deal with the interstellar extinction and with transient nuclear burning on a white dwarf issues later in the book, but the atmospheric extinction is discussed here.

The question of atmospheric extinction making Sirius appear red has been addressed by a number of authors. Three different approaches have been considered: (a) a single observation of Sirius by See (1927), (b) a systematic measurement of reddening of stars when close to the horizon performed from Mallorca (Graff 1931), and (c) an extensive discussion of astronomical extinction in the Mediterranean region (Evans 1987).

The atmospheric extinction is the result of the influence of air molecules, of small dust particles floating in the atmosphere, and of water droplets as well as other types of aerosols, on the propagation of the light rays from the observed object through the atmosphere to the observer. The molecules and aerosols in the atmosphere modify the light coming from an object by absorbing parts of it and by scattering other fractions out of the beam into different directions. This, by the way, is the reason the day sky is blue; the blue light emitted by the Sun scatters more than does its red light. The Sun looks redder at sunrise and at sunset because its light passes through more air (a longer path through the lower atmosphere), thus suffers more scattering and loses a larger fraction of its blue component.

The influence of the extinction is, therefore, a function of the amount of air between the observer and the object. The length of the light path, for a plane-parallel atmosphere of thickness h, is

$$l = h \times sec\,z \qquad (3.1)$$

Here z is the zenith distance and this formula is an approximation valid for $z < 75°$. The way this atmospheric thickness depends on the zenith distance is called the *air mass* X. Under the approximation given above,

$$X = sec\,z \qquad (3.2)$$

For considering the visibility of celestial objects close to the horizon, or for a discussion of the heliacal rising of Sirius, this approximation that

TABLE 3.3. Zenith distance and airmass evaluations

z	$\sec z$	X_{min}	X_{ave}	X_{max}
80.0	5.76	5.55	5.60	5.62
82.0	7.19	6.79	6.88	6.91
85.0	11.47	10.13	10.39	10.50
87.0	19.11	14.68	15.37	15.68
87.5	22.93	16.44	17.35	17.77
88.0	28.65	18.60	19.83	20.42
88.5	38.20	21.29	23.01	23.84
89.0	57.30	24.73	27.16	28.37
89.5	114.59	29.18	32.72	34.54
90.0	∞	35.09	40.39	43.23

the extinction is simply proportional to sec z is not acceptable. An exact treatment for the air mass in a molecular Rayleigh-scattering atmosphere was given by Kocifaj (1996). The optical air mass is a function not only of the zenith distance, but also of the amount of water vapor and dust in the atmosphere. Kocifaj calculated maximal, average, and minimal air masses for various apparent zenith distances. A comparison of his values and those calculated from the simple $sec\,z$ approximation is shown in Table 3.3.

The three values listed for Kocifaj (1996) represent minimal (column 3), typical (column 4), and maximal (column 5) values of the airmass for different zenith angular distances. The three values for each z describe the range of change expected mainly from the variation in air density at the Earth's surface, and by the vertical gradient of the air density. It is possible that meteorological factors also influence the airmass, but Kocifaj did not take these into account. The tabulated values (listed in Table 3.3) agree reasonably well with those calculated by Schaefer (1993) in his excellent paper on the limits of astronomical vision.

The influence of aerosols in hindering the visibility of Sirius at heliacal rising was hinted to already by Schoch (1924), who found that the dates of heliacal rising in antiquity were different from Babylon to Alexandria. From this, he concluded that Alexandria should have had a constant layer of mist (tiny water droplets with radii between 1 and 20 μm) covering the lower part of the horizon, even when the rest of the sky was clear. If this would have been the case, then Sirius would mainly have been dimmed, without being significantly reddened. This is because the small water droplets of the mist produce mostly "grey", i.e., wavelength-independent extinction, as shown by e.g., Casperson (1977). However, the heliacal rising of Sirius in connection with the Nile flooding was probably observed by the ancient

Egyptians in the vicinity of Memphis near present-day Cairo, not from Alexandria, and the issue of horizon fog would not be valid. On the other hand, if Ptolemy did observe the heliacal rising of Sirius from the terrace of the Alexandria library, sea-shore morning fog would have been a relatively common phenomenon.

A modified explanation for the redness of Sirius involving extinction was put forward by Whittet (1999). He calculated that some kind of atmospheric extinction, while Sirius is extremely low in the sky, could make it red without making it extremely faint. Whittet's reasoning uses atmospheric extinction by molecules to redden Sirius, instead of aerosol particles or water droplets in mist. The solid particles are relatively large in comparison with the wavelength of optical light, and produce primarily "grey extinction", as do water droplets. The air molecules, on the other hand, produce scattering that is very strongly wavelength-dependent. This "Rayleigh" scattering produces extinction that behaves as

$$A_\lambda \propto \lambda^{-4} \qquad (3.3)$$

where λ is the wavelength of the affected light. The goal of the calculation, as Whittet explained, was to make Sirius red without making it too dim. This is because the brightness of Sirius should always be comparable with that of the other bright stars, in particular with Arcturus, Betelgeuse, Aldebaran, Antares, and Pollux, so that it would remain one of the brightest objects in the sky.

The atmospheric extinction is expressed by the astronomers as a series of extinction coefficients for the standard photometric bands. It is possible to translate the individual extinction coefficients in B (k_B) and V (k_V) into a reddening coefficient at zenith E'(B–V)=k_B–k_V, and from this into a total-to-selective extinction ratio R=$k_V/E'(B - V)$. Whittet (1999) used the extinction coefficients for the Sutherland site of the South African Astronomical Observatory (SAAO) to obtain R≈1.25; such an extinction ratio would redden Sirius appreciably without dimming it too much relative to the bright stars. At the Wise Observatory in Israel, presumably more representative of the Mediterranean climate where Ptolemy and his predecessors made their observations than the SAAO, k_B ≈0.20 and k_V ≈0.14 (Brosch 1992), yielding R≈ 0.14/0.20 − 0.14 = 2.33. Nawar (1999) measured the extinction for the Kottamia observatory, half-way between Cairo City and the Suez canal, in autumn 1982. His values are k_B = 0.41 and k_V = 0.28, from which one derives $R \simeq 2.15$. Both values are significantly different from that used by Whittet.

Assuming the typical atmospheric conditions for the Wise Observatory, requires that Sirius would be at a rather high airmass ($\sim$16), which would imply a zenith distance of 87°.5 (Kocifaj 1996). In order to make Sirius into a moderately red star one would need to redden its B–V≈0.0 color by about one magnitude. At such a low altitude, Sirius would have been dimmed by ~2.3 mag, turning it into a V≈0.8 mag object. In addition, the scintillation and the differential refraction would have been very significant. In principle, therefore, the proposal by Whittet could be a reasonable explanation of the perceived redness of Sirius since with this kind of extinction the star does not become extremely dim while it becomes significantly red. However, one should consider that this had to take place only at the time of heliacal rising, above the background of the rising Sun and in absence of horizon haze or fog. This modifies somewhat the observing conditions and changes the results. Locations free of fog, haze, etc. are often encountered on mountain tops (such as those where astronomical observatories are now located). In such cases, Rayleigh scattering as a main extinction contributor would be valid. However, neither Cairo nor Alexandria are mountain sites, thus it is unlikely that Whittet's explanation would be valid.

The issue of the visibility of Sirius during the heliacal rising was addressed by Schaefer (2000). He also provided routines to calculate heliacal rising of celestial bodies (Schaefer 1985) and for estimating visibility limits in difficult situations, such as heliacal rising, when the star is reddened, its light is extinguished by the atmosphere, and it must compete against the increasing sky brightness due to the rising Sun (Schaefer 1998) in order to be visible. Schaefer (2000) found that in order for Sirius to be visible at heliacal rising, its altitude must be ~6° while the Sun must be ~5° below the horizon.

An altitude of 6° puts the air mass of Sirius, when it first becomes visible at heliacal rising, at 8.7–8.9 (using the data of Kocifaj 1996) or a bit more than 10 (with Schaefer's data from Table 3 of his 1993 paper), where Sirius will be dimmed by 1.2 mag only, appearing as a ~ 0th mag star (still one of the brightest in the sky), but with a B–V color of only ~0.5 (orange or yellowish, and not very red). Using the average Kocifaj airmass for the heliacal rising from the table, still with the Wise Observatory extinction coefficients, we find a dimming of Sirius to $m_V \approx 0$ and a color B–V≈ 0.6 for an altitude of 5° above the horizon. This indicates that atmospheric reddening at heliacal rising might be an acceptable explanation, even if one could somehow explain why people do not call Sirius a red star even today. However, this requires the use of the extinction coefficients derived by Whittet (1999) that are very unlikely for the sites used to observe Sirius in ancient Egypt.

The idea of a red Sirius at heliacal rising was supported by Brecher (1979), where he quoted the Arab poet Ibn Alraqqa:

> *I recognize Sirius shining red, whilst the morning is becoming white.*
> *The night, fading away, has risen and left him.*
> *The night is not afraid to lose him, since he follows her.*

At this point it is valuable to consider the personal experience of Holberg (2007) who observed Sirius near the horizon from Kitt Peak, a 2000-m high astronomical observatory. Holberg wrote that he never saw a red Sirius except once at setting, when he used binoculars so that he could perceive the color of the star even when it was extremely low and so exceedingly extinguished that it could not be seen with the naked eye. This testimony, from a very experienced astronomer, puts a serious question mark on all attempts to explain a red Sirius by atmospheric extinction, even if this is caused by Rayleigh scattering.

If one rejects atmospheric extinction as a possible explanation, since it would have turned Sirius into a red star only under very extreme conditions and at very specific and rare instances while leaving it appear white most of the time, one has to resort to astrophysical causes to explain the redness. One such intriguing possibility was already suggested by Brecher (1976). He proposed that Sirius B could be a very hot white dwarf star, with an effective temperature $\sim 200,000°$K and thus very young, and that it was once a red giant that lost its envelope in historical times. This proposal is not tenable now, since the effective temperature of Sirius B was measured accurately, as will be discussed below, and it is much cooler (thus the star is much older) than Brecher's value. Another explanation, by D'Antona & Mazzitelli (1978), was that an event of Hydrogen burning in a thin shell accreted by a white dwarf, lasting a few thousands of years, could have been responsible for the redness. This will also be discussed later. Among other things, they caused me and Dr. Itzhak Nevo to search for an ejected stellar envelope near Sirius that could manifest itself as an emission nebulosity (Brosch & Nevo 1978). The frustration connected with our negative results was certainly one of the triggers to write this book.

However, before dealing with astrophysical issues it is necessary to address other mysteries connected with Sirius; its binary nature and the amazingly detailed knowledge of it by the Dogon tribe in central Africa.

3.3. The binary nature of Sirius

At present, Sirius is an outstanding celestial object. Considering the immediate Solar neighborhood, α Canis Majoris is the brightest star by far. Figure 3.3 shows the color-magnitude diagram of the brighter stars of CMa. Sirius is the topmost object to the left, brighter by more than 3 mag from the next bright star.

Data about the bright stars (m$\leq$6.5), which can be seen with the naked eye provided the observations take place from a dark site, are collected in the Bright Stars Catalog (Hoffleit & Jaschek 1982). The photometric information extracted from this source and limited to stars within the

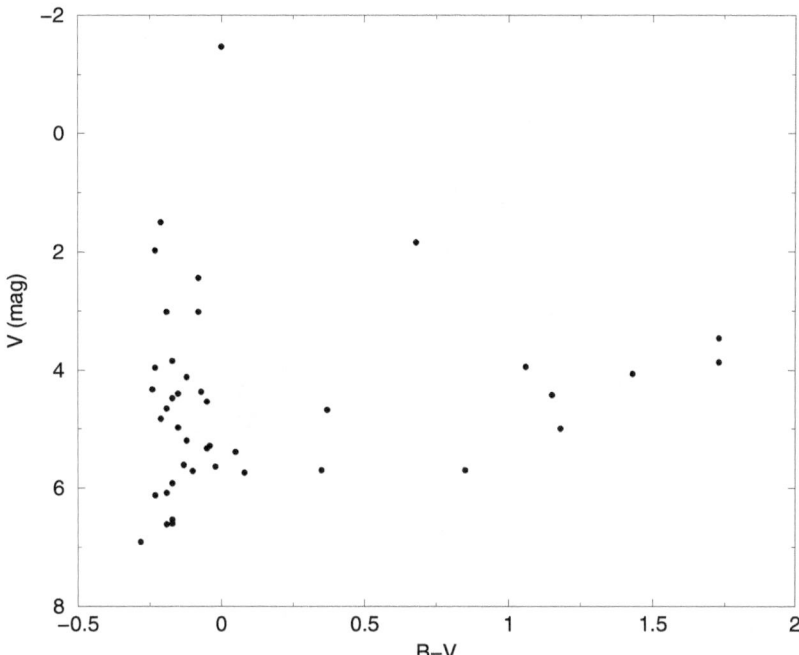

Figure 3.3.   The relatively bright stars in the Canis Majoris region of the sky are all rather blue, more than the usual percentage in a randomly selected sky area. Data from the 4th edition of the Bright Stars Catalog (Hoffleit & Jaschek 1982).

constellation Canis Majoris, which is plotted in Figure 3.3, emphasize the brightness of Sirius in comparison with its neighbor stars. The point at the top of the diagram is Sirius, well separated from the clearly defined main-sequence. The B–V color of Sirius is bluer than that of the Sun and one can, therefore, conclude that Sirius is hotter in its outer regions that are visible to us than is the Sun.

Most stars in the immediate Solar vicinity are rather old. Figure 3.4 shows the distribution of the nearby stars, within 25 pc from the Sun, in a color-magnitude (CM) diagram. The sequence of cooling white dwarfs is clearly visible, as are the sub-dwarfs and a sprinkle of giants or stars on their way to become giants. This separation of white dwarfs, main sequence stars, and red giants is already visible in photometry obtained more than half a century ago. Sirius does not stand out in this diagram, since it is neither very blue or very red, nor does it have an exceedingly bright intrinsic magnitude.

A color-magnitude diagram plotted by Eggen (1950) for stars within 14 pc (his Figure 1) shows quite well the main-sequence and some stars that seem to belong to a horizontal branch. The CM diagram, and the evolution of stars with specific application to the system of Sirius, will be discussed in a later part of this book.

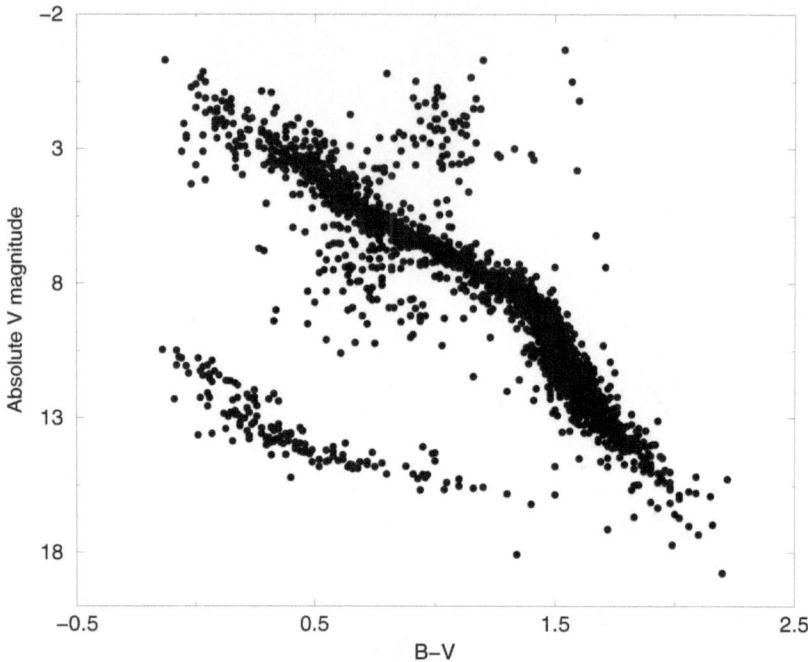

Figure 3.4. Color-magnitude diagram [M$_V$ vs. (B–V)] for stars listed in Gliese's Catalog of nearby stars (third edition) that have their B–V color listed. Sirius is located near the upper-left corner of the plot, but it is not the extreme point.

Now it is clear why Sirius is such a bright object; primarily this is because this star is close to us thus it appears bright, and also because it is intrinsically brighter than most stars. Sirius was not always the bright blue star we see these days in our Southern sky, because, relative to the Sun and to the Earth orbiting it, Sirius is moving. About five million years ago it was in the constellation Lynx, more than 70° from the position it now has, and appeared much fainter. A good graphical description of its motion was published by Lovi (1989), following an article published in the Dutch astronomical yearbook *Sterrengids* for 1989. This diagram was specially reworked with up-to-date parameters for Sirius by Dr. Rob van Gent for this book (Figure 3.6).

    The apparent change of position of Sirius on the sky, as seen from the Solar System, is the combination of a number of motions. First, α CMa has its own motion in the Galaxy; so does the Sun and the entire Solar system. This causes both the location of the object we look at, and the location we look from, to change with time. We are aware only of the relative change of position for Sirius with respect to the "fixed" stars that, in reality, are much more distant than Sirius, therefore their apparent motions are small to negligible. The combination of both motions is the "proper motion" and

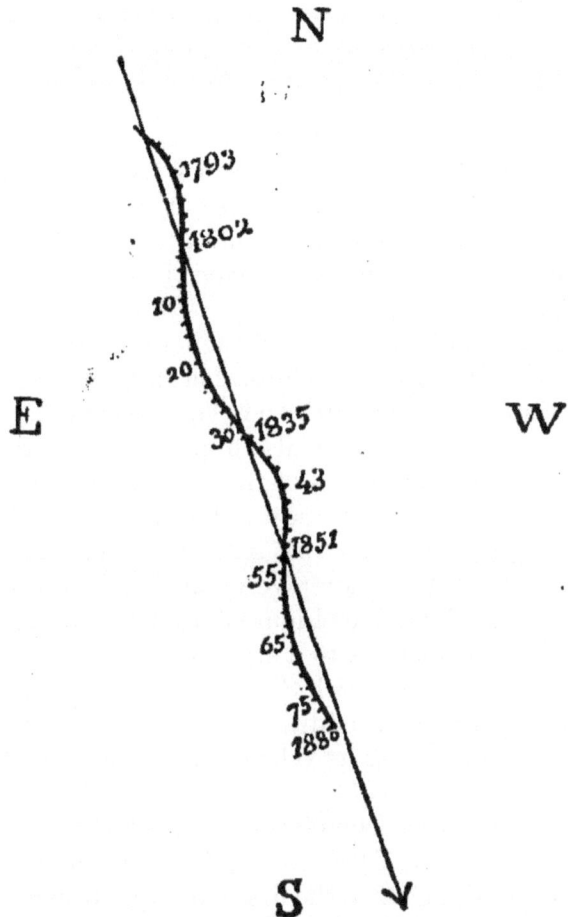

Figure 3.5.   The apparent motion of Sirius for about one century, as plotted in 1884 by Flammarion.

is represents a change in the relative location of Sirius. Its units are arcsec per century for most stars, or arcsec per year for the nearmost, fast-moving stars ($10".3 \, \mathrm{yr}^{-1}$ for Barnard's star).

The chart in Figure 3.5 represents the proper motion of Sirius over ~100 years and is reproduced from Flammarion (1884), who mentions that the other bright stars in the general vicinity of Sirius hardly change position in such an astronomically-short time interval. I will revisit this point below, when modern determinations of the location of stars by the *Hipparcos* astrometry satellite shall be discussed.

The proper motion of Sirius was discovered by Halley when he was studying the positions of bright stars. Halley published a short paper in the *Philosophical Transactions* (Halley 1717) where he wrote that

these stars [Sirius and Arcturus] being the most conspicuous in Heavens,
are in all probability nearest to the Earth, and if they have any particular
Motion of their own, it is most likely to be perceived in them....

Halley discovered the movement of Sirius among the "fixed" stars during his study of the rate of precession of the equinoxes. By comparing the measured position of Sirius in relation to other stars with that derived from ancient catalogs, Halley found a discrepancy of 37 arcmin (Porchon 1899), more than half a degree, which is significant, given the estimated accuracy of the observations.

The proper motion of Sirius is one of the largest on the sky: $1".34$ yr^{-1}, as measured by the *Hipparcos* satellite. The largest proper motion of a star belongs to Barnard's star ($10".3$ yr^{-1}), a nearby 9.6 mag M4V star located 1.8 pc from the Sun that is also approaching us with a velocity of 106.8 km s^{-1}. The two components of the proper motion for Sirius, in the direction of the Right Ascension and in the direction of the Declination, are $\mu_\alpha = -0".5461$ yr^{-1} and $\mu_\delta = -1".2231$ yr^{-1}. This implies that Sirius is moving to lower Right Ascension (West) and Declination (South) values, as Flammarion's figure shows (see Figure 3.5). The total proper motion, at the distance of Sirius, translates to a transverse space velocity of $\sim$15 km s^{-1}. This is the velocity with which Sirius moves perpendicular the line of sight as seen from the Sun; in addition, Sirius has a radial velocity with respect to the Sun representing its velocity along the line of sight; this will be discussed later.

The motion within the nearby Galaxy causes Sirius to "close in" to the Sun; five million years ago α CMa was a faint star, barely visible at $\sim$ 6*th* mag. During this period, in which the first man-apes became Homo Sapiens, Sirius came much closer to the Sun and became a –1.5 mag object, i.e., a thousand times brighter. Sirius became the brightest star in our skies only $\sim$90,000 years ago, when it overtook Canopus in brilliance. It will maintain this title for about 210,000 years while continuing to brighten slightly from its –1.44 mag for the next $\sim$60,000 years as it draws nearer the Sun. At its closest it will be only a bit further than eight light-years away in the constellation Columba (the Pigeon, south of Lepus the Hare) and it will shine as a –1.7 mag star.

At this point, and considering the perceived motion of Sirius among the stars, it is interesting to mention a proposition by Gore (1903) based on one of the Arabic names of Sirius: *al-schira al-abur* ("Sirius which has passed across"). Gore linked this name with Al Sufi's mention of a mythological explanation that Sirius crossed the Milky Way in the direction of Canopus. This obviously happened during the Stone Age and, if correctly representing the human memory of an astronomical event, is truly amazing.

Sirius has two other motions. One is a well-known "wiggly" displacement, which is the manifestation of the revolution of Sirius A and Sirius B around their common center of mass. The binary character of Sirius will be discussed extensively later. This modulation of the proper motion of Sirius was noticed for the first time by Bessel[19] in 1834, at a time when the existence of Sirius B was not known. Since Könisgberg is located at 54°43' North, Sirius would have been seen there at most 20° above the Southern horizon, and its celestial position would have been heavily affected by refraction. For this reason, Bessel limited his investigations to positional changes in Right Ascension.

The discovery was communicated to the astronomical community by Herschel (Bessel 1844) as the translation of part of a letter from Bessel to him. Bessel wrote:

...existing observations entitle us without hesitation to affirm that the proper motions, of Procyon in declination, and of Sirius in right ascension, are not constant; but on the contrary, that they have, since the year 1755, been very sensibly altered.

Bessel concluded that the two stars must have "invisible" companions. The gravitational attraction of the invisible companion forces an orbital motion of the primary (brighter) star; this would be super-posed on the general pattern of the proper motion producing the wide wiggles with a period of about 50 years. Bessel wrote

There remains then the explanation (2) alone. Stars whose motions, since 1755, have shewn remarkable changes, must (if the change cannot be proven to be independent of gravity) be parts of smaller systems. If we were to regard Sirius and Procyon as double stars, the change in their motions would not surprise us; we should acknowledge them as necessary, and have only to investigate their amount by observation. But light is no real property of mass. The existence of numberless visible stars can prove nothing against the existence of numberless invisible ones.

Interestingly, almost a century before Bessel's discovery of the binary nature of Sirius, Voltaire[20] already described it as having a companion in his 1752 *Micromegas*.

Voltaire wrote in the opening paragraph of *Micromegas*:

Dans une de ces planètes qui tourne autour de l'étoile Sirius,...

(on one of the planets that circle around the star Sirius).

One wonders whether this proves a special, arcane knowledge of Sirius B, or even of a planetary system around Sirius, on the part of Voltaire. Very

---

[19]Friedrich Wilhelm Bessel (1784–1846), German astronomer and mathematician, directed the Königsberg observatory from 1810 to 1846.
[20]François-Marie Arouet (1694–1778), French writer, essayist, deist and philosopher, known by the pen name Voltaire.

The Path of Sirius from −5 Ma to +5 Ma

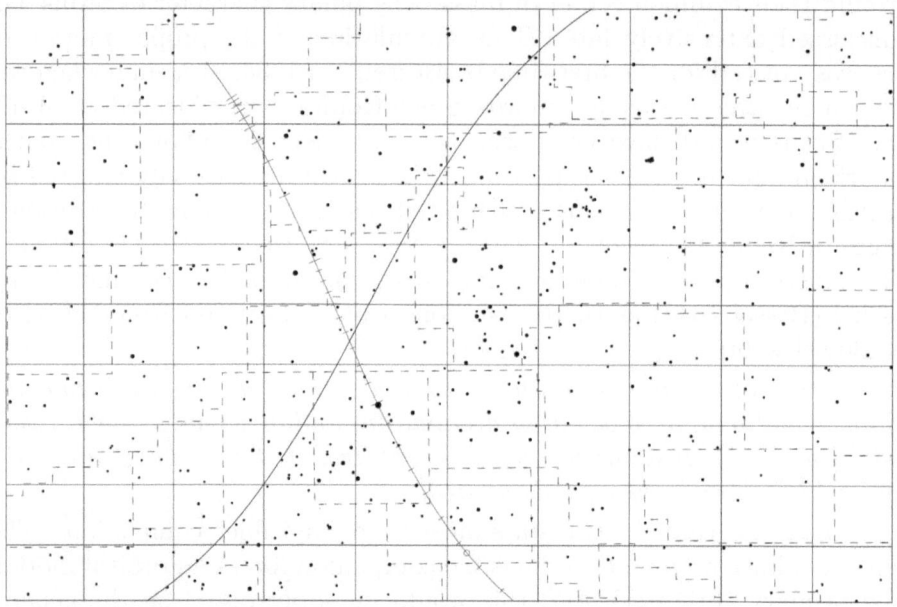

Figure 3.6. The apparent motion of Sirius among the stars since 5,000,000 years ago (courtesy Dr. R. Van Gent). The tick marks are plotted every 10^5 yrs for $\pm 10^6$ yrs, but for times $\pm 10^5$ years near the present they are plotted every 10,000 years. The IAU borders of the constellations are plotted together with stars brighter than 5*th* mag.

likely the answer to this is negative. As other Encyclopaedists of his time, Voltaire believed in the plurality of worlds and of life. This is probably why he decided to locate one of his protagonists on a planet orbiting the brightest star in the sky, while the other one, a dwarf in comparison with the humongous Micromegas who was eight miles tall (*" huit lieues, vignt-quatre mille pas géometriques de cinq pieds chacun"*), was an inhabitant of Saturn.

One more reference in *Micromegas* is thought-provoking, in connection with the material already presented; in the second chapter Voltaire reveals that the sun of Micromegas has a reddish tinge (*" Notre soleil tire sur le rouge"*); did Voltaire indeed intend to imply that there is a third body in the system, apart from Sirius A and B and was the sun of Micromegas' planet, and that this body is red? Did he refer to claims of redness of Sirius (as will be discussed below) or to a red Sirius? Was Voltaire the inspiration for Barker's article that was published only eight years following Micromegas?

Coming back to the indications of binarity for Sirius, Bessel did not live to see his prophetic discovery confirmed since he died in 1846. This task was relegated to his follower to the directorship of the Königsberg observatory[21], Peters. He combined all the observations up to 1848.6 and solved the orbit of Sirius A and B: these two bodies revolved about each other in an eccentric orbit with a period of 50.093 years (Peters 1851). The importance of Peter's work, where he also published the predicted locations of the second component of the system, was emphasized by Holberg (2007); these predicted future locations allowed directed searches for the optical counterpart.

The contribution of Peters was soon superseded; Safford (1862) and Auwers (1864) refined his results by including in the analysis the positional perturbations in Declination and showed definitively that Sirius is a double star, even though the secondary had not yet been observed. The detection of Sirius B took place in 1862 and will be described below.

The trajectory described by the two components of Sirius shows an additional component of motion as tiny wiggles super-posed on the wider modulations of the proper motion by the orbital motion of the Sirius pair. The wiggles, with a period of one year, are a reflection of the motion of our vantage sight-point, the Earth, on the apparent position of the Sirius system. Our observatories orbit the Sun with the planet, with a period of one year and on an orbit which is $\sim$300 million km wide. This means that our location oscillates during one year around the Sun with the orbital period of the Earth and changes from one side to the other across the ellipse (which is very close to being a circle) after 6 months. We are not aware of this motion because we participate in it, along with all our "local" reference points, but accurate astronomical observations are able to detect it.

The Earth's orbital motion around the Sun gives rise to annual parallax. Since our location changes around the Sun on an ellipse, the nearby stars, when projected against the more distant objects, appear to move on tiny ellipses. In the case of Sirius, this additional motion is super-posed on the:

1. Orbital motion of the pair
2. Proper motion of the barycenter of the system
3. Reflection of the proper motion of the Solar system

The parallax of the Sirius system is so very large ($\sim$0".38) that it makes the size of the tiny wiggles noticeable. The wiggles of the parallactic motion imposed on the track of Sirius among the stars are not exactly sinusoidal; this is because of the relative inclination of the Earth's orbit to the line of sight to α CMa. In about 60,000 years, when Sirius will reach its closest location to the Sun on its celestial path, the apparent amplitude of the parallax will be largest. However the trajectory will not be as close to a sinusoidal curve as one could imagine; this happened $\sim$300,000 years ago when Sirius was close to the ecliptic plane.

---

[21]Christian August Friedrich Peters (1806–1880) was the German astronomer who followed Bessel as director of the Königsberg observatory.

3.4. The Dogon tribe and a modern Sirius mystery

An interesting question related to Sirius is the advanced knowledge held by the West African Dogon tribe of intricate details about the α CMa system. This could not be expected and is believed by some to be a prime indicator of a major cultural and technological transfer, by extra-terrestrial beings and many thousands of years ago, to the human civilization. However, as will be explained below, one should not accept this as a ready explanation. Instead, the evidence points indeed to a cultural transfer, but much closer to out times and not from extraterrestrials.

The Dogon people live on a large plateau in the Homburi (or Hombori) mountains near Timbuktu, a region in Mali south of the Sahara desert in North Africa and have complicated and colorful rituals (Figures 3.7 and 3.8). According to Dogon oral tradition, the tribe settled in this area between the fourteenth and the fifteenth centuries. Perhaps the most important of their religious teachings, as reported by sociologists and relevant to the present discussion, is their knowledge about a star that is invisible to the eye in the Sirius system. Some claim that the Dogon received their knowledge by visitors to the Earth from another star system and, indeed, this is one of the corner stones of New Age beliefs in Sirius.

The most extensive collection of facts, myths, and plain suppositions connecting the Dogon and Sirius can be found in the book *The Sirius Mystery* (Temple 1987). Briefly, the thesis proposed there is that extraterrestri-

Figure 3.7. Dogon ritual mask dance (Photograph by Dr. Galen R. Frysinger, used with his permission).

Figure 3.8.   Another Dogon ritual mask dance (Photograph by Dr. Galen R. Frysinger, used with his permission).

als from the Sirius system landed on Earth and triggered the phenomenal advances of Mediterranean civilizations since ~3,000 BCE. This proposition was summarized also in a short paper by Temple (1975).

Temple based his thesis on a description of the beliefs of the Dogon tribe and of a few related tribes, collected by Marcel Griaule (1898–1956) and Germaine Dieterlen (1903–1999), two anthropologists who studied the Dogon from the late 1930s to the 1950s. Following Griaule's death, Dieterlen remained active in African studies.

Temple read the summary of the Griaule and Dieterlen's findings in a book about African mythology to which Griaule and Dieterlen contributed an article (which had been badly translated into English). Following Part One of his book, Temple (1987) included that entire Griaule and Dieterlen article, after its translation into English had been rechecked by a professional translator. I refer to this article as GD, with the understanding that it is the translation published in Temple's 1987 edition of his book, with the attached footnotes.

One of the Dogon rites concerning Sirius is related to their calendar. Every ~60 years, so the Dogon believe, the world is renewed. This happens at a ceremony called Sigui, which was heralded by a rock fault in the Yougou Sigui village turning red, and by a certain type of gourds appearing in a spot outside the village. The elders recognize these 60 year periods by counting 32 years intervals, each celebrated with a drinking bout and by the deposition of a cowrie shell within the rock fault. Records of the masks used in this ceremony indicate that this it has been held since around the thirteenth century.

In their renewal rites, GD claim that the Dogon make extensive reference to Sirius and its companion. They call α CMa A by the name *sigi tolo* (*sigo bolo* in the related Bambara tribe), or *yasigi tolo*=the star of Yasigui. The focus of the ceremony, however, is the tiny companion of Sirius, which they call *põ tolo*. The word *tolo* means "star" and *põ* is the smallest seed the Dogon know of. GD noted that the seed corresponds to the plant with the botanical name *Digitaria coilis* and called the star *põ tolo* by the name Digitaria.

Digitaria, according to the Dogon, is extremely dense:

> It consists of a metal called sagala, which is a little brighter than iron and so heavy that all earthly beings combined cannot lift it.

A tongue-in-cheek reckoning of this density can be based on the calculated mass, 3.8×10^{10} gr according to GD, and diameter "the size of a stretched ox-skin", say 2×10^2 cm diameter. This yields a density of $\rho = 9 \times 10^3$ gr cm^{-3}, only three orders of magnitude off the typical average density of a white dwarf! A modern mass estimate yields a density even closer to that of a white dwarf. Assuming that one average person can lift a 20-kg mass, and that there are six billion people on Earth, the density of the lifted mass should be at least 3×10^7 gr cm^{-3}.

The Dogon believe that Digitaria orbits Sirius on a ~50 year orbit. Even so, though they say that *põ tolo* circles the star Sirius every 50 years, the special Sigui renewal ceremony takes place every 60 years, to celebrate the completion of one cycle associated with *põ tolo*. The elliptical orbit, shown schematically in Figure 3.9 brings Digitaria close to Sirius, where it shines brighter, then far from it when Digitaria shows a twinkling effect. This could be interpreted as implying that Digitaria does not have its own light but only reflects the light of *sigi tolo*=Sirius, as would a planet do.

The orbit presented here, drawn after *Figure iii* in DG, can be compared with the actual projected orbit of α CMa B around its primary apparently showing a measure of amazing correspondence. But Pesch & Pesch (1977) objected to this interpretation and pointed out that the ellipse should not be

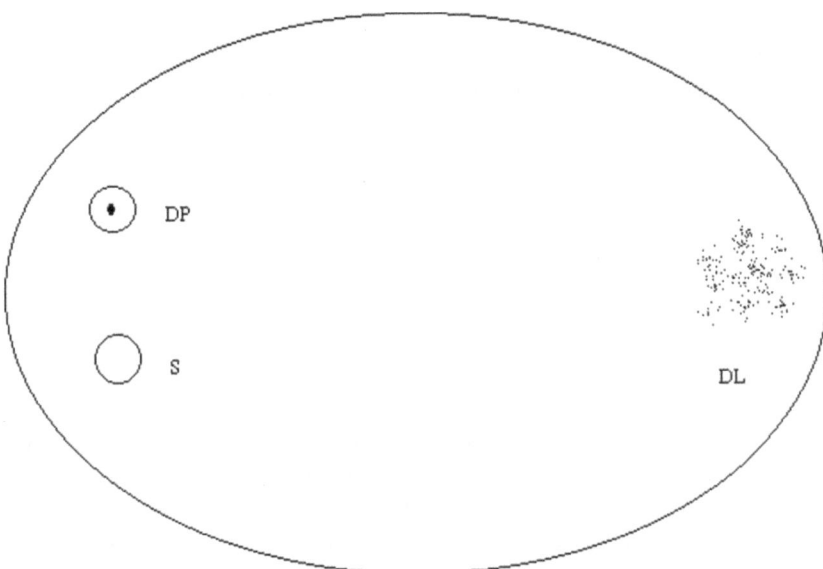

Figure 3.9.   The trajectory of the star Digitaria around Sirius (after Figure *iii* in GD). Two positions of Digitaria are marked, one near Sirius (DP) and another far from it (DL).

taken to mean the orbit of Sirius B, but rather representing the boundaries of the drawing. Apparently, many of the drawings produced by the Dogon have egg-like boundaries.

A side mention in GD has not been remarked on before. The second sentence of the part *The Origins and Features of Digitaria*, in the translation given in Temple (1987) on p. 42 of his book, is:

Sirius appears red to the eye, Digitaria white.

It is not clear whether GD actually quoted words from the mouth of their informant. It is tempting to believe that the Dogon knew something about the presumed color change of Sirius, though why would they choose to believe a star to be red when it was so obviously white at the time the interviews took place is hard to understand. One possibility is that they believed most of the light coming from the system to originate from Digitaria and to be modified according to its distance from *sigi tolo*.

Another weird claim made by the Dogon and reported by GD is that the Sirius system actually consists of three stars. The third is called *emme ya*="sorghum female", it is four times lighter than Digitaria and circles around it

along a greater trajectory and in the same time as it

i.e., also ~50 years. Accepting the identification of Digitaria as Sirius B, *emme ya* would then be its satellite and it, together with Digitaria, would travel together circling Sirius A. Other objects in the Sirius system are also mentioned by the Dogon, such as the "Star of Women", the *Yourougou*, etc. As will be explained below, a third body in the Sirius system was reported and was searched for intensively, starting from the second decade of the twentieth century, but none was found.

In addition to their knowledge of Sirius B, the Dogon mythology includes accepted facts such as Saturn having rings and Jupiter having four moons. These features cannot be seen with the naked eye by the vast majority of people. The Dogon have four calendars, for the Sun, Moon, Sirius, and Venus, and have long known that planets orbit the sun. In the description of GD, the extensive astronomical collection of beliefs held by the Dogon appears impressive.

Irrespective of the claim of Temple (1987) that the celestial knowledge of the Dogon was passed on to them as heritage of an ancient Mediter-ranean civilization, advanced beyond others by extraterrestrial visits, some attempts were made to rationalize this knowledge as a version of the "cargo cults" encountered in Polynesia after contact of the natives with American soldiers during the second World War. Cargo cults are cultural transfers from an advanced culture to a less advanced one, by which the teachings become (part of) the basic philosophy of the less advanced culture. These cults, and a possible connection with the Sirius cosmogony of the Dogon, were discussed by e.g., Brecher (1979).

Brecher evaluated the Dogon contacts with Europeans as extremely limited, at least until the early years of the twentieth century. He was con-vinced that Griaule and Dieterlen were scrupulous researchers and did not fabricate their data. Moreover, he mentioned that Griaule was practically a member of the Dogon. Following Griaule's death *"about a quarter of a million Dogon gathered and held a state funeral"* showing how appreciated he was by the members of the tribe. Brecher concluded that the Dogon indeed showed some knowledge of the Sirius system that was unknown to humankind (if Temple's "extraterrestrial" hypothesis is rejected) at least until 1862. He hypothesized that a chance contact with an educated Jesuit priest, perhaps in the 1920s, brought about the astronomical knowledge to the Dogon and its subsequent incorporation into their basic beliefs. In support of this he mentioned that references to Sirius B could be found in widely distributed newspapers of the 1920s such as the New York Times, the Scientific American, and Le Monde.

Temple (1981)wrote in his reply to criticism by Carl Sagan:

Long ago I wrote to the head of the White Fathers in Mali, Father Dubreuil at Mopti, who replied saying that none of their missionaries had had any contact with the Dogon before 1949.

The missionary society known as "White Fathers" (*Pères Blancs* in French), is a Roman Catholic Society of Apostolic Life founded in 1868 with the purpose of conversion of the Arabs and of the peoples of Central Africa to Catholicism. From Temple's letter, it follows that the White Fathers are not Jesuits as hypothesized by Brecher (1979), and that there is no evidence that other missionaries did not have contact with the Dogon.

The issue of a binary Sirius system in the Dogon mythology could be strengthened by the fascination of the Dogon with twin-ness (Pesch & Pesch 1977). This reference also offers an explanation for the cultural transfer of astronomical knowledge to the Dogon: the French schools operating there since 1907 and offering geography and natural sciences in their curricula. Whether this could be the factor that transferred some modern astronomical knowledge to the Dogon or not is hard to determine. It is clear that most high-school graduates of today, even those who live in developed countries, are unaware of any of the knowledge about the Sirius system the Dogon were supposed to have.

Obviously, any solid evidence of an early contact by the Dogon with some astronomical authority (presumably from some Western civilization) that took place before their being investigated by the French anthropologists, would strengthen an idea of cultural transfer as the source of their knowledge about Sirius.

One possible "smoking gun" that could identify the source of the astronomical knowledge of the Dogon comes from an essay on Sirius written by Herrmann (1988). In a note to the second edition of his book he mentions an account of the French expedition led by Henri Deslandres to observe the total solar eclipse of April 16, 1893 from Central West Africa. The track of that eclipse started on the coast of Chile, passed over Argentina, Paraguay and Brazil, then crossed the Atlantic and reached the African coast North of the river Gambia ending in the Sahara. The most successful eclipse observations of this event were obtained in South America (Chambers 1912, pp. 150–151), where French, British, and American teams were deployed.

The eclipse of April 16, 1893 was memorable because Schaeberle[22] discovered and reported a comet (Comet I 1893, Schaeberle 1894a, b, c) on the negatives that show the eclipsed Sun. The comet was first discovered on the plates exposed in Chile and, subsequently, identified also on the plates received from Brazil and from Africa. The closest angular distance reported between the comet and the Moon's limb, on April 16, was 0.86 Solar diameters. Cliver (1989) re-evaluated this comet discovery report and concluded that what was really observed was not a comet, but a disconnected coronal mass ejection.

---

[22]John Martin Schaeberle (1853–1924) was a German–American astronomer.

The appendix to Harrmann's book (Herrmann 1988) mentions an account of the French eclipse expedition to West Africa, which was written by a young Romanian astronomer named Coculesco or Coculescu[23]. Coculescu completed his university studies at Bucarest's Faculty of Sciences and obtained his Mathematics diploma in March 1889. A year later, in December 1890, he left for Paris to train further at the Observatoire de Paris. There, in 1892, he published his first work on celestial mechanics. In 1893, while still stationed in Paris, Coculescu participated in the African expedition headed by Henri Deslandres in order to observe the solar eclipse.

Herrmann (1988) mentions in the appendix to the second edition of his book, basing his description on the account of Coculescu, that the eclipse expedition stayed in the field for 5 weeks; during such a long period many contacts with the native population were presumably made, some perhaps with members of the Dogon tribe in attendance. It is reasonable to assume that a large scientific expedition, geared for astronomical observations, passed on the then-accurate knowledge of astronomy to the population by stories and by night-time viewing of the sky through their telescopes. The local inhabitants might have been exposed to quite a terrifying experience, with the Sun being eaten out of the sky right in front of their own eyes. Some of the teaching of the European "superior beings" equipped with shiny brass tubes that allowed one to see very far by day, and displayed such amazing views as the rings of Saturn and the moons of Jupiter by night, could then have been incorporated in their beliefs.

This "smoking gun" identified by Herrmann (1988) provides, therefore, a natural mode by which some astronomical knowledge of Sirius could have been passed on to the Dogon, but does not confirm it. One strengthening argument for this possibility would be to see whether the information which could have been passed on to the Dogon in 1893, on the nature of the system, corresponds to what they believed about it. The nature of α CMa B being a faint and massive object had been known since Bessel's dynamical investigation in 1844, and the companion star Sirius B was actually seen already in 1862 by members of the Clark family and, following the discovery of Sirius B, by many other astronomers. The issue of a possible color change of Sirius in historical times was publicized in France by Flammarion (1884) and is understandable that Coculesco would have been familiar with the contents of that book. The one unexplained item, for which there is yet no "smoking gun", is the claim of the Dogon for the existence of a third component in the α CMa system; such claims were apparently first published only in the 1920s and could not have been passed-on to the Dogon by members of the 1893 eclipse expedition.

---

[23]Nicolaie Coculescu (1866–1952), Romanian astronomer and professor at the Bucharest University.

However, the story of Sirius and the Dogon does not end here. Another the Belgian anthropologist Walter van Beek followed in the footsteps of Marcel Griaule and studied the Dogon for a decade starting on January 1978 (van Beek 1991). His research found a number of disturbing discrepancies with the findings reported by Griaule and understood to be at the core of the Dogon culture. One of these was that in the 1980s the Dogon did not seem to be aware of Sirius as a double star or as a triple star, as asserted by DG in *Le Renard Pâle*[24]. The Dogon studied by van Beek considered Sirius to be just a star, *dana tolo* (the hunter's star), and no knowledge of stars was required for daily life or in ritual. In particular, the informants interviewed by van Beek never heard of *sigu tolo*, or *po tolo*, or *èmè ya tolo*, the names of Sirius A and B reported by Griaule.

van Beek (1991) described Griaule as having a *strong personality, with firm convictions and clear preferences.* Apparently, Griaule worked with a small number of informants during his numerous short field trips, a single principal informant and a number of secondary ones, and had only a limited command of the Dogon language. He appeared to the Dogon as *a figure of slightly more than human proportions, with an uncanny insight into the hidden thoughts and motives of people.* The behavior exhibited by Griaule during his research is characterized by van Beek as a result of Griaule's belief that the Dogon culture is limitless and all-knowing.

van Beek (1991) went on to the subject of Sirius near the end of his paper. One of his informants, Amadingué, who worked with Griaule's chief informant Ongnonlou and consulted with him, reported that the informant never spoke of Sirius as a double-star system. He described Sirius as the grandfather in a "father-and-son" relationship with nearby stars (two other stars in Canis Majoris). The inference that the informant spoke of Sirius as a binary would, therefore, be the result of Griaule's imagination. In fact, the remainder of the circle of informants used by Griaule did not agree about which star is meant by *sigu tolo*: it could be a star whose rising would announce the *sigu*, or Venus, itself. They all agreed, however, that they learned all about this star from Griaule himself.

The description of the disturbing ethnographical study of the Dogon by van Beek (1991) puts at least one additional question mark on the possibility that the Dogon had some previous knowledge of the binarity of Sirius and of the companion being a white dwarf prior to their contacts with the nineteenth and twentieth century astronomers and anthropologists. The explanation put forward, that the cultural transfer was due to Griaule himself, perhaps involuntarily, makes sense also because the

---

[24] "The Pale Fox", the name of the book about the Dogon written by Griaule and Dieterlen and published in 1965.

period when Griaule was active, from the 1930s, was also the time when a third star in the Sirius system was reported a number of times by the astronomers.

However, first-hand evidence from Griale's daughter, Dr. Geneviève Calame-Griaule, specifies that Marcel Griaule was not aware of the significance of the Sirius stories he collected from the Dogon until he returned to Paris. Quoting from her reply to van Beek's (1991) article (Calame-Griaule 1991):

> As for his alleged training in astronomy, I can report that his training was in literature; he had no notion at all of astronomy, and it was the Dogon who first began telling him about the stars. If he later displayed charts of the heavens, it was for his own use and not to instruct the Dogon. As for the satellite of Sirius, he was completely ignorant of its existence until the Dogon spoke to him of a "companion", at which point he consulted the astronomers of the Paris Observatories and found them as surprised as he was.

It is also possible that Griaule himself changed from the time he published his first Dogon study in 1938 to the publication of *Le Renard Pâle* in 1965. Preston Blier wrote in a comment to van Beek's paper that *Le Renard Pâle* is more abstract and philosophical than Griaule's early, empirical work.

The historical perspective of a possible third body in the Sirius system will be presented below. Knowledge of this could have enhanced and enriched the earlier cultural transfer that presumably took place in 1893.

3.5. Conclusions

For some unknown reason, Sirius was perceived as a red star, at least from a few centuries before the current era until about the end of the first millennium. This perception was possibly localized to the European-Mediterranean region.

The historical evidence seems often too vague to derive any firm conclusion about a physical meaning of possible color changes among the stars and, in particular, for Sirius. Color evidence is unlike other types of astronomical records, such as solar or lunar eclipses, planetary alignments, apparitions of novae and supernovae, comets, etc., which are very useful in disentangling astrophysical phenomena and in timing of historical events.

Despite this vagueness, it is also not possible to fully discount the possibility of a red Sirius based only on the historical and cultural arguments. It is useful to seek physical explanations for such a possibility, since these might reveal interesting and hidden facts about Sirius itself. The particular explanation involving atmospheric extinction at heliacal rising seems contrived and not fully believable.

The strange measured motion of Sirius against the background of stars was explained by Bessel as the influence of an unseen companion. The thorough investigation of position vs. time of Sirius allowed the prediction of the location where such a companion might be found.

The extensive knowledge of the Dogon tribe regarding the Sirius system could be explained mostly by infrequent contacts with astronomers or with other persons with modern knowledge about the astrophysical properties of Sirius. Instead of finding in the Dogon myths a trace of celestial influence, or of an intervention of extraterrestrial beings, we possibly encounter here another example of an experimental result influenced by the experimenter; in this case, the anthropologist collecting the evidence.

Chapter 4
Approaching modern times

In the previous chapter it was explained that Sirius is a double star and that there were claims that it is actually a triple system. Here the issue of multiplicity will be discussed in detail.

4.1. The discovery of Sirius B: a tale of gravity

The prediction of an unseen companion to Sirius is due, as described above, to Bessel in 1844, and the actual discovery to Clark on January 31 of 1862. However, the first "guess" that such a body does exist was published more than 100 years before the actual discovery. Voltaire, in his book *Micromegas* published in 1752 and which is an imitation of Swift's *Gulliver's Travels*, put the hero on an immense satellite of the star Sirius, as explained above. However, this was very probably only a lucky guess on the part of Voltaire.

Bessel noticed already in 1834 that the proper motion of Sirius, the change in its location relative to a background frame of reference, was variable. In 1840, Bessel noticed a similar behavior in the proper motion of Procyon (α Canis Minoris). Bessel concluded that this indicated the presence of invisible companions in double systems. He wrote to Humboldt (quoted from Aitken 1964, p. 237):

> *I adhere to the conviction that Procyon and Sirius are genuine binary systems, each consisting of a visible and an invisible star. We have no reason to suppose that luminosity is a necessary property of cosmic bodies. The visibility of countless stars is no argument against the invisibility of others.*

Bessel's precise observations and careful calculations allowed him to deduce that the perturbing but invisible companion orbits Sirius with a 50-year period. Bessel was also able to predict its position relative to the bright star, a factor that would weigh heavily in recognizing and accepting the 1862 visual discovery by the Clark family, as will be described below.

Bessel died in 1846 and the urge to find the hidden companions of the stars with variable proper motion waned. The first theoretical orbit of

Sirius B was calculated by Peters and triggered a search for it at a number of observatories. The discovery, however, was a combination of luck and technology.

The Clark family, starting with Alvan Clark, were telescope makers for half a century. Alvan was originally an engraver, then an inventor, and only after his son George Bassett Clark built a telescope in 1844, (that he equipped with a bronze mirror), was Alvan Clark convinced to start building telescopes. The optics firm Alvan Clark & Sons was founded in 1851 and became soon well known for the high quality lenses it was producing. These lenses went into the largest telescopes then known, including an 18.5-inch refractor originally ordered by the University of Mississippi, but which was not delivered due to the Civil War and the lens remained in the Clark workshop.

The discovery of the secondary, by Clark in 1862, was reported in a rather dry manner by Bond (1862) in the scientific press:

An interesting discovery of a companion to Sirius was made on the evening of Jan. 31 by Mr. Clark with his new object glass of eighteen and a half inch aperture. I have been able to observe it with our refractor of fifteen inches, as follows:

1862, Feb. 10, Angle of Position   85° 15±1° 1
*   Distance   10".37±0".2* *When the images are*
tranquil the companion is distinctly enough seen, but these moments are quite rare, as the low altitude of Sirius exposes it to almost continual atmospheric disturbances.

The story, as told by Flammarion (1884), is slightly different and a bit more exciting. He writes that when Alvan Clark was about to finish working on the 47-cm diameter lens, his son Alvan Jr. tested the lens on Sirius. Wishing to check the color correction, he pointed the telescope to a bright and blue star, Sirius. The son suddenly exclaimed "Father, Sirius has a companion !". This was immediately confirmed by Alvan Clark himself and the following day it was communicated to the Harvard College Observatory, whose director was Bond.

The discovery of the companion of Sirius was quickly confirmed at a number of observatories on both sides of the Atlantic. The companion, Sirius B, has also been colloquially known as "The Pup", since it is the faint companion to the The Dog Star.

Two questions had to be answered positively before the nature of this object could be determined: (1) was it really a companion of Sirius, or just a nearby star seen in projection, and (2) did it follow the path predicted by the gravitational calculations. At the beginning, it seemed that the faint star might just be in the background. Otto Struve, the director of the Pulkovo Observatory, could not fit the existing observations to the predicted orbit (Struve 1864). He rejected the assumption that the faint star was in the background, writing

> ...*William Herschel, at the end of the last century–when with this supposition, the small star ought to have had a distance of about one minute from Sirius–never has noticed its existence though it is well known that about that time he has frequently observed Sirius...*

The large proper motion of Sirius implied that at the time Herschel observed the system a background star would have been quite distant from the glare of Sirius, thus relatively easy to discern. Struve attributed the misfit of the orbit to observational errors, since from Pulkovo (near St. Petersburg) Sirius would always have been observed at low altitude, thus its image would be affected by scintillation and differential refraction.

However, a few years of additional observation convinced Struve that the newly discovered star was indeed the true companion of Siris. He went even further and estimated that its mass was about half that of the bright star, Sirius A (Struve 1866), noting that it was between 8*th* and 10*th* magnitude thus about 10,000 times fainter that A. From this and the mass he concluded that

> ...*both bodies are of a very different physical constitution.*

The position of the faint companion star near Sirius, being just as predicted theoretically so as to account for the perturbations in the proper motion of Sirius, helped the astronomical community accept the reality of the companion.

As is probably the case in many discoveries, there have been claims for an even earlier sighting of Sirius B, prior to the Alvan Clark Jr. observation. One such claim was described by Botley (1959). She quoted a series of sources, ending with the report from March 1861, 10 months before the discovery of the Clark family, of a *small shining point* near one of the diffraction rays emerging from Sirius A. However, even if true, this was not followed up by the professional astronomers.

The location of α CMa B, measured at the time of its discovery, was almost exactly as that predicted by the gravitational theory that accounted for the "wiggles" of Sirius A. Aitken (1964) quotes an investigation by Safford in 1861, based on meridian circle positions of Sirius A, which *"assigned to the companion a position angle of 83°.8 for the epoch of 1862.1"*. An observation by Bond for the epoch 1862.19 placed the B component at 10.07 arcsec from the primary and in position angle 84°.6.

Since its discovery and until the close of the twentieth century Sirius B circled its primary slightly less than three times. The positions of the two components of the binary system have been measured time and again. The orbit of the B component around A, shown in Figure 4.1, is known

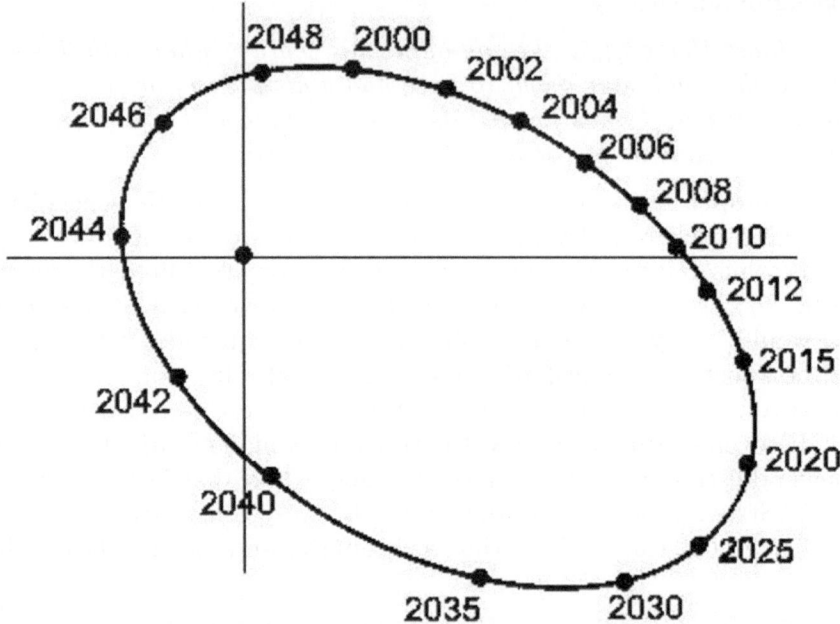

Figure 4.1.   The orbit of Sirius B around Sirius A, with the location of the faint star plotted for a number of years in the first half of the twenty-first century. The angular size of the semi-major axis of the orbit is 7".5. North is down and East is to the right.

now with reasonably high precision, despite the difficult and challenging conditions of observing a faint star very near a bright one.

It is worthwhile to discuss these observational difficulties, in order to appreciate the difficulty facing the early observers. When the Clark family discovered Sirius B, the distance between the two stars was somewhat more than 10 seconds of arc. This is a small angular separation, equivalent to the perceived diameter of a 5 cm disk at a distance of 1 km, but one that can be resolved with reasonable instruments. The problem is the relative brightness of the two objects, since Sirius A is ∼10,000 times brighter than Sirius B.

When the two bodies are close together, and the angular distance between them is only a bit more than three seconds of arc, it is very difficult to separate them in order to measure the relative positions with reasonable accuracy. Following the discovery of Sirius B by Clark, the Sirius system was followed for almost 30 years, until the spring of 1890, when Burnham measured a separation of 4".19. Burnham (1891) reviewed all the available observations of Sirius B since its discovery and concluded that the period should be of order 53 years and that the closest separation between the A

and B components should take place in 1892–1895. He revised the period
to 51.97 year (Burnham 1893), following the correction of a mistake caught
by See.

The following measurement was taken only four years later, in 1896, by
Aitken, when the separation was 3".81 and when Sirius B was already past
its peri-astron (the peri-astron passage occurred in 1894). The additional
observations after the peri-astron passage allowed, together with the previous observations since the discovery of Sirius B, a glimpse at 255° of the
orbit (Burnham 1897), confirming its parameters.

The time-series of positions allowed an excellent definition of the positions of the system stars with time. The parameters of the orbits in the
Sirius system were thus well-defined and are explained by the diagram in
Figure 4.2. Aitken (1964) quoted for the epoch of 1894.02 an eccentricity
e=0.588, an apparent semi-major axis a=7".62, an argument of peri-astron[1]
$\omega = 145°.87$, an inclination of the orbital plane[2] i=+44°.49, and a position
angle of the nodal point[3] $\Omega = 43°.77$ (quoted from Volet 1931). The period
given there is 49.94 years.

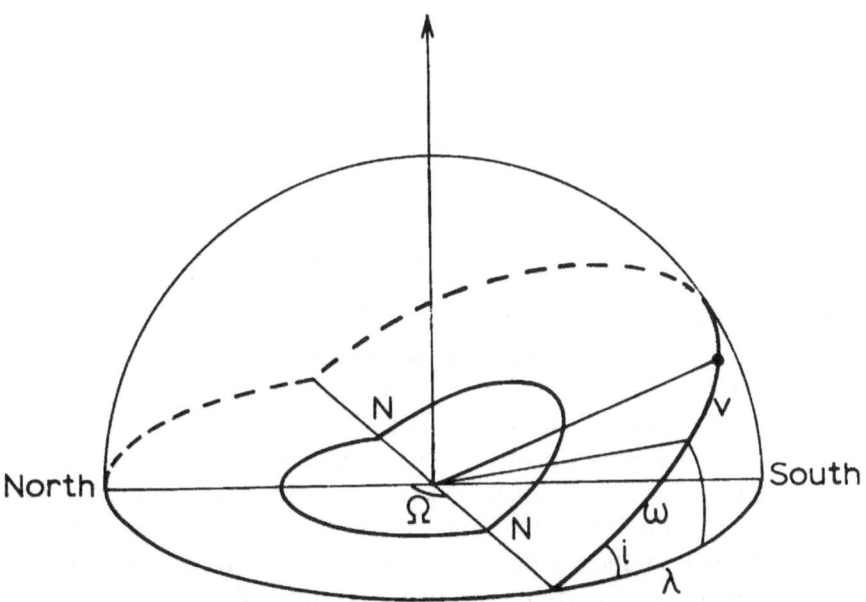

Figure 4.2.   Relevant parameters that define the orbit of a binary star, after Minnaert
1969. The vertical arrow points to the observer (Earth).

[1]The angle in the plane of the true orbit between the line of nodes and the major axis.
[2]The angle between the orbit plane and the plane perpendicular to the line of sight.
[3]The position angle of the line of intersection of the orbit plane with the plane perpendicular to the line of sight.

Given that the orbit is well-known, it is relatively easy to calculate the combined mass of the two stars. Assume that the orbit is seen pole-on, that is, the orbital plane is perpendicular to the line of sight. From Newton's formulation of the third law of Kepler, a simplified relationship between the average distance between the two stars and the period of the orbit is:

$$\frac{a^3}{P^2} = \frac{G(M_1 + M_2)}{4\pi}. \tag{4.1}$$

If the system is viewed in a pole-on orientation, and if we know the distance to the system from the accurate measurement of its parallax, we can convert the average distance from angular units to astronomical units. Note that the early parallax measurements for Sirius were not very accurate: Henderson (1839) reported a value of 0.23 arcsec with an error smaller than 0.25 arcsec, thus a parallax consistent with zero. Gill (1898) observed from Cape Town and obtained a much more accurate value of 0.370±0.007 arcsec. If the orbital period P is measured in years, the distance between the components a in astronomical units, and the sum of masses in solar units ($M_\odot$), (4.1) transforms into the sum of masses:

$$M_1 + M_2 = \frac{a^3}{P^2}. \tag{4.2}$$

Inserting the values for Sirius we find $M_1 + M_2 \approx 3.2$ $M_\odot$. It is possible to split the sum of masses into individual masses by using, for example, the average distances of each star from the barycenter, the center of mass of the system (Sirius A is about twice closer to the barycenter than Sirius B).

Aitken (1964) gave a total system mass of 3.36 $M_\odot$ and a mass ratio of 0.39. This made the secondary component an object with $M_B \approx 0.94$ $M_\odot$, that is, nearly the mass of the Sun. Modern values have $M_A \approx 2.14$ $M_\odot$ and $M_B \approx 1.05$ $M_\odot$. The latest values, based partly on unpublished work, are $M_A = 2.02 \pm 0.03$ $M_\odot$ and $M_B = 1.00 \pm 0.01$ $M_\odot$ (Holberg 2007).

The relative (then absolute) magnitudes were even more difficult to measure than the relative positions. Even when the two stars were well-separated, the exceeding brightness of the A component made the visibility of the secondary difficult. Knowledge of the brightness of the secondary is necessary in order to allow a first estimate of its nature, given that its mass is known from the orbital dynamics of the system. When the two stars are close together, it is next to impossible to measure the light coming from the faint component. The reason is that in any optical system there is scattered light, which comes mainly from the bright star. Its primary origin, assuming perfect lenses and mirrors, is the diffraction pattern created by structures in the optical path, such as the "spider" holding the secondary mirror in a two-mirror telescope. Any grain of dust on the optics adds to this, by scattering some of the light into arbitrary directions, including toward the focal plane instruments.

Measuring the light from the α CMa system posed, therefore, a serious challenge to nineteenth century astronomers. While it is not difficult to measure the light from the entire system, since this is contributed mainly by the bright A component, the ratio of intensities of the two stars is very difficult to measure. Early measurements used double-image systems, whereby a small thin prism would be inserted in the sky beam producing a second image of the field, strongly attenuated relative to the main image produced by most of the field that was not affected by the thin prism element. The reason is that a thin prism causes mostly a deviation of the light rays and not so much dispersion of the different colors. The entire field is "shifted" laterally with respect to the center of the main field of view. The ratio of the main and secondary image intensities equals the ratio of the surfaces presented to the sky by the unobstructed part of the telescope field and by the diverting prism. This way, it is possible to project an attenuated secondary image of Sirius A that can be compared with the primary image of Sirius B, thus finding the ratio between their light intensities.

A second method to measure the light from Sirius B is by using an objective-grating arrangement. This device is made of a set of narrow, parallel and equidistant opaque strips mounted at the entrance aperture of the telescope, which produce a diffraction pattern of the light. In this case, as in the thin prism arrangement, secondary images form in the image plane, given that the number of strips in the grating is small and their spacing is wide. It is possible to arrange the spacing of the secondary images by adjusting the distance between the strips, and change the direction of the line of images by rotating the grating device.

At this point it is interesting to notice that a first indication of of the color of Sirius B was published soon after the discovery of the optical companion. Knott (1866) observed the companion with a 7.5-inch Clark refractor and remarked that

> Its color was a fine pale blue (about Blue[3] of the Late Admiral Smyth's chromatic scale...

Admiral Smyth[4] devised a scale of color specifically for binary stars and published it in 1864 as a booklet called *Sidereal Chromatics*. This was a first attempt to systematize that what we now call "effective temperature" of stars. This blue color is an indication of the high effective temperature of the faint Sirius companion.

The two methods of measuring the relative intensities of Sirius A and B, double-image prisms or objective gratings, were used in the nineteenth

---

[4]Admiral William Henry Smyth (1788–1865) was a noted hydrographer and astronomer

Figure 4.3. Sirius A and B can be seen in this image obtained with a digital camera on the Mt. Wilson 60-inch telescope. Note the cross pattern of the diffraction caused by the supports of the focal assembly at the prime focus, as well as smaller spikes from other obstructions in the light path. The distance between the two components of the binary at the time this image was obtained was about 5 arcsec and Sirius B is the tiny image to the left of the bright star (Credit: Jimmy Westlake, Colorado Mountain College, reproduced with permission).

century and in the first half of the twentieth century, and yielded a photometric magnitude of –1.6 for Sirius A and ∼8.5 mag for Sirius B. In the early 1930s Kuiper (1932) measured the companion to be 8.44 mag. A repeated measurement by Kuiper (1934), using the 12-inch refractor of the Lick Observatory, yielded a value of 8.42 mag. Both values were obtained with coarse gratings in front of the telescope objective. On the other hand, at about the same time Vyssotsky (1933) measured 7.1 mag with the McCormick Observatory's 26-inch refractor using a rotating sector photometer; this value, however, was considered to be an outlier among the entire set of measurements.

Given that the distance to the Sirius system was known from a measurement of its parallax, this implied an absolute magnitude of 11.3 for α CMa B. This is ∼ 540× fainter than the Sun, while the mass of the star is comparable to the mass of the Sun. Sirius B belongs, therefore, to the class of objects classified as "white dwarfs", small and compact objects that are not normal, main-sequence stars.

Even more difficult than the measurements, were scientific interpretations of the existence of such a faint object near Sirius. One early interpretation, by Hussey (1896), was that Sirius B shines by reflecting the light of Sirius A as a kind of giant planet. Although he mentioned that such an assumption is doubtful, Hussey nevertheless used it to calculate the light curve at different phases. This is because at that time very compact stars, such as white dwarfs (WDs), were not known.

A similar but more modern photometry method was described by Lippincott & Worth (1966). It was developed to allow observations of the Sirius system with the Sproul 24-inch refractor using photographic plates. The seeing at this observatory was of order 2 arcsec and the image of Sirius A was so large as to make its size approximately 8 arcsec in radius! The observers devised a hexagonal aperture for the refractor; this caused the diffraction pattern to change from that of a circular aperture. A hexagonal-shaped entrance aperture produces images with six bright spokes radiating away from the bright star. The six-spoked pattern used by Lippincott & Worth could be rotated, by rotating the hexagonal aperture at the sky end of the telescope with respect to the sky. The result of using a hexagonal objective aperture, in the proper orientation, is that one can arrange to have the image of Sirius B fall between the bright spokes of the image of Sirius A and thus become measurable.

Searching for the white dwarf companion would have been much easier in modern times. Barstow et al. (2001) used the Hubble Space Telescope and the Wide-Field/Planetary Camera to image stars known from space-based observations to have hot white dwarf companions. Ground-based observations could not separate the WDs, but their presence was inferred from the detection of a UV or Extreme UV excess (using the *ROSAT* Wide-Field Camera or the *EUVE* satellite; see below). The detection of a hot WD in a binary system was eased by obtaining the image in the UV; this enhanced considerably the detectability because in this spectral range the relative contribution of the WD in a binary system is larger. HST images in the UV also have a much better resolution than anything presently obtainable from the ground. This allowed Barstow et al. to detect WDs in eight of the 17 systems they surveyed. The use of modern tools to investigate the Sirius system will be discussed later.

4.2. A third body in the Sirius system?

The first claim that a third body might exist in the Sirius system was published by Jonckheere (1918), where the orbit of the binary system in α CMa was thoroughly investigated. His study was based on a series of positional measurements ranging from 1862.19 to 1917.18. Jonckheere mentioned that

> a small increase in the distance [between A and B] in 1908...and the
> curious run of the residuals in these last observations [from 1894.10]
> have led me to suspect the existence of a perturbing body [in the system]
> (my additions in square brackets).

Jonckheere continued to observe the Sirius system with the Marseille 26-cm telescope in the following decade and reported the position of the B component (Jonckheere 1930), but did not remark again on a possible third component.

In the late 1920s, a number of experienced observers reported the presence of a third body in the Sirius system. For example, Fox (1925), observing from the Dearborn Observatory, reported for 1920.110 (February 9) that Sirius B

> appears persistently double in [position angle=]231°: [separation=]0".8
> (my additions in square brackets).

On February 4, 1926 Innes and others at the Union Observatory in South Africa saw a third star in the Sirius system. This was reported as a 12th magnitude object, 1.5–2 arcsec away from the 8th magnitude Sirius B.

van den Bos (1929) wrote that the C component was not visible at the Union Observatory in Johannesburg on January 15, 1929, but that it was seen later that year together with Sirius B:

> On 1929.199 at PA=135°.9 and separation 1".7 from Sirius B, and on
> 1929.227 at PA=129°.2 and separation 1".38 from B.

van den Bos mentioned that the C component is "all doubtful", but gave its magnitude as 3.0 and 2.5 for the two occasions when he saw it (he listed consistently a magnitude of 7.5–8.0 for the B component). Finsen (1929) mentioned six observations of the putative C component, also performed from Johannesburg in 1928–9, given here in the format of (year, PA, separation, "comment"):

> (1928.219, 68°.6, 1".83), (1928.221, 77°.6,–, "weak, bad definition"),
> (1929.041, "no trace"), (1929.167, 129°.4, 1".38, "C sharp and stel-
> lar, regarded as quite certain, good measure"), (1929.199, 126°.4, 1".5e),
> (1929.227, 128°.0, 1".25e).

He mentioned that the observation on 1929.041, when the C component was not seen, was done while the filar micrometer wires were suspected of having been fiddled with. Only for the last measurement Finsen listed a magnitude for the C component: 3.0 mag. This, by the way, is consistent with the magnitudes reported by van den Bos (1929) from the same observatory.

Finsen's extensive note for the observation on 1929.227 is worth quoting:

for this measure a bar about 2 inches wide was fixed in front of the objective in position angle 55°±, giving a remarkable dark lane in the out-of-focus light surrounding Sirius A. This dark lane extended almost to the image of A on both sides and had the appearance of opposed two parabolas with Sirius A touching the vertex of each. B and C, lying in the lane, were consequently much more easily observable. This device may prove useful in other cases where a bright star has a faint relatively close companion (e.g., Procyon). The real existence of Sirius C is still regarded as doubtful.

Note though that previous experienced observers, who inspected visually the Sirius system before the first reports about the existence of a third body in the system were published, did not report any such possibility (e.g, reports by Holden 1896 and by Hussey 1896). However, 1919 was the year when Sirius A and B were most distant and the detection of a third component would have been easiest. Before 1919, involvement in the First World War might have influenced the attention of the observers. After ~1940, the approach of Sirius B to peri-astron would have made the observation of Sirius C rather difficult.

The various observations of Sirius C in the 1920s, and the apparent disturbances of the orbits of Sirius A and B, have been studied first by Zagar (1932) and by Volet (1932). These two studies checked specifically the residuals of the motion of Sirius B around the A component, and searched for a periodical perturbation of B that could explain the deviations from the calculated two-body solution. Both studies concluded that a period of 6.40–6.42 years, and a motion with an amplitude of 0".14 and low eccentricity, almost co-planar with the motion of B around A, could fit the residuals. The motion of C around B had to be in an opposite sense to the motion of B around A. Unfortunately, the positions of Sirius C, as measured by van der Bos and by Finsen and quoted above, did not correspond to the predictions of the orbits of Volet and Zagar.

With the calculated orbit, the derivation of some physical parameters for the third body became possible. Volet (1932) derived a mass ratio $\frac{C}{B} = \frac{1}{16}$. Adopting this mass ratio for the present accepted mass of Sirius B (~ 1 M$_\odot$) yields a mass for the C component of ~ 0.06 M$_\odot$, just below the Hydrogen burning limit, implying that the C component could be a "brown dwarf", according to presently-accepted limits.

A theoretical possibility arguing for the presence of a third body in the Sirius system, specifically a white dwarf companion to Sirius B, was apparently proposed by G.P. Kuiper (cf. Marshak 1940, p. 352). This was in order to account for a discrepancy between the measured value for the radius of the white dwarf and a theoretically calculated value. As quoted by Marshak, Kuiper concluded that:

> *the ad hoc assumption of a close, faint companion to Sirius B, which is at the same time, a white dwarf, could bring μ_e[5] to the value of 2.*

Negative evidence for the existence of a possible C companion was mentioned by Aitken (1964). He testified that he never saw such a companion, and neither did Burnham or Barnard, both experienced observers of double stars known for their keenness of vision. Another piece of negative evidence was by Hetzler (1935), who studied the brightness of Sirius B in a number of (photographic) bands. He wrote:

> *No visible evidence of a third star has been found on the photographic or the red plates. If it is within one magnitude of B in brightness, the separation of any such star would probably be less than a second of arc.*

A series of observations specifically designed to attempt a detection of Sirius C was initiated by Lindenblad (1973). He used the US Naval Observatory 24-inch refractor to obtain multiple-exposure images of the Sirius system and measured positions of the A and B components over a period of 6.8 years, from 1965 to 1972. The experiment was specifically designed to detect a positional disturbance with a period of 6.4 years, as proposed by Zagar (1932) and Volet (1932). The results indicated a need to revise the orbit determined for the Sirius A-B system, but did not reveal traces of periodic perturbations that could be the signature of a close third companion.

A similar lack of periodic modulation of the orbit was reported by Gatewood & Gatewood (1978). Their study covered a 60-year period, starting with first-epoch photographic exposures from the Yerkes Observatory obtained from 1917 to 1923, and reaching until 1977 with plates from the same observatory. Based on these results, Gatewood & Gatewood redetermined the parallax of the system and refined the orbit parameters by reducing the errors in the orbit parameters.

Other results relevant to the possible presence of a third low-mass companion in the Sirius system originate from theoretical studies of stable orbits of third bodies in binary systems. A general analysis of such third-body orbits was performed by Dvorak (1986). He separated the orbits into "satellite" and "planetary" types. The first kind of companions orbits very close to one of the binary components; the second orbits essentially the center of mass of the binary system, very far from the two stars. It is possible that in the Sirius system this could be the situation, and it is worth-while to consider Dvorak's results.

---

[5]Mean molecular weight per electron, a parameter used in models of stellar structure.

Benest (1989) performed numerical simulations of this problem of three-body system stability and found that a stable orbit with a period of 6.635 years (around Sirius A), and one of 0.69 years around Sirius B, can be the largest allowed such orbits.

Stable orbits in the system of Sirius have also been considered by Holman & Wiegert (1999). This was done in the context of a number of binary systems in elliptical orbits, and the stable orbits were calculated by numerical integration of test particles for one billion years. Holman & Wiegert found that the inner stable (satellite) orbits could have periods of 2.2 years around α CMa A and 1.76 years around B, and that the "planetary" orbit that was stable had a period of 398 years for a semi-major axis of 79 a.u. Obviously, even wider orbits would also be stable, but very wide ones could be disturbed by random encounters between the Sirius system and field stars.

4.3. Modern searches for a third companion

The more recent searches for a possible third body were done using more modern observational techniques. An interesting novel method was used by Bonnet-Bidaud & Gry (1991) to image the immediate neighborhood of Sirius. They used the 1.5-m Danish telescope at ESO and imaged Sirius and its neighbors with a CCD camera, after they attached a small plastic cone to the entrance window of the CCD. This small cone acted as a coronagraph mask, blocking light from a region ~25 arcsec in diameter from entering the CCD detector. In addition, they modified the diffraction pattern of the telescope by adding a mask with eight circular apertures to the entrance of the telescope. The images were obtained in the U, B, V, I and Gz (Gunn-z) photometric bands.

Bonnet-Bidaud & Gry (1991) presented an image of a 2'.5×4' field centered on Sirius. The image shows nine stars for which they give a location relative to Sirius, and their photometric measurements. The photometry is not very accurate, only good to 0.3 mag, because of the strong remaining background of scattered light produced by Sirius. Table 4.1 reproduces their results.

The projected distances from Sirius, given in the second and third columns, refer to the date of observation (1985). The columns V_{B-V} and V_{V-I} give the visual magnitude calculated using different color terms [this is not explained in the paper but can be guessed when examining the table]. The last column of Table 4.1 gives the absolute magnitude the stars would have, if their distance from the Sun would have been the same as the distance to Sirius. Two objects, no. 4 and to a lesser extent no. 3, have very red colors. They could be very late-type stars that, given their absolute

TABLE 4.1. Photometry of stars in the immediate neighborhood of Sirius
(from Bonnet-Bidaud & Gry 1991)

Star	$\Delta\alpha^s$	$\Delta\delta"$	V_{B-V}	B–V	V–I	V_{V-I}	M_V if at 2.7 pc
1	−35.41	79.82	14.2	1.3	2.0	14.2	17.1
2	32.33	55.65	14.5	0.9	1.5	14.6	17.4
3	50.45	−35.71	16.4	2.0	2.5	16.4	19.2
4	47.67	−60.33	17.2	3.0	2.8	17.2	20.1
5	18.91	97.1	16.6	0.3	2.1	16.6	19.4
6	−78.93	86.78	16.8	1.1	2.3	16.9	19.7
7	13.64	−69.91	–	–	2.3	17.6	20.4
8	−17.23	92.02	–	–	2.2	17.9	20.7
9	−35.47	104.55	–	–	2.0	17.8	20.6

magnitudes in the last column, might be physical companions in the Sirius
system.

Benest & Duvent (1995) re-analyzed the orbit of Sirius A and B and
found evidence for a periodic disturbance with a 6.05 year period and an
exceedingly small amplitude, of only 0.055 arcsec. Benest & Duvent suggested that Sirius could indeed be a triple star. The third companion could
have a stable orbit around Sirius A, with a period of about 6 year, if it
had a low enough mass (≤ 0.05 M$_\odot$). A comparison of these results with a
numerical simulation of a triple system like that of Sirius indicated that an
orbit with a 6 year period around Sirius A would be stable, but one around
Sirius B would not.

A search for a third body in the Sirius system was performed by Kuchner
& Brown (2000). They used the HST near-infrared camera NICMOS to
search for dust close to nearby stars, specifically in the systems of Sirius,
Altair, and Procyon. The observations had to reduce significantly the light
from the bright star in order to allow a sensitive search for scattered light
from possible dust disks. Therefore, they were performed with the NICMOS
camera 2 in coronagraphic mode, with the F110 filter that isolates a spectral
band at 1.104 μm.

As a by-product of this search, Kuchner & Brown (2000) were able to
measure the near-IR flux from Sirius B (well-resolved from the A component
at 1.1 μm and at the HST resolution) and to set stringent limits on the
possibility of a faint third body in the Sirius system. The near-IR flux
measured for Sirius B was 0.503 ± 0.15 Jy.

The limits for a faint third body, derived from the near-IR images of
the Sirius system, were so sensitive that they could rule out the existence
of a body as faint as an M-dwarf that could exist farther than 1.8 arcsec

from Sirius A and of a star hotter than an L-dwarf[6] farther than ~3 arcsec from Sirius A. The volume of space very near Sirius A could not be completely surveyed. In particular, any low-mass companion closer than 4.2 a.u. from Sirius A could not have been detected, because it would have been located in the saturated part of the image. This is the orbiting distance for a companion of Sirius A in a 6-year period. However, due to the faintness of Sirius B and its distance from the A component at the time the HST measurements were performed (3.79 arcsec in October 20 and 22, 1999), it was only possible to rule out companions of Sirius B that would be as bright as brown dwarfs in a 6-year period.

Another negative search for a possible third companion was reported by Schroeder et al. (2000). They used the HST Planetary Camera with the F1042M (1 μm) and F814W (I-band) filters to search for companions by directly imaging the field. The existence of the very bright A component of the system required the determination of its image center by interpolating to the mid-line of the diffraction spikes (the accuracy achieved was $\pm$0.2 pixel), determining the position of Sirius B by subtracting a bi-variate polynomial for the local background, and searching for any residual image. The position of α CMa B was found to lag by ~75 days from that calculated from the orbital elements given by Benest & Duvent (1995). This indicates that a revision of the orbital parameters is in order. Note that the search method used by Schroeder et al. does not rule out the presence of a companion that would be fainter than M6V and would orbit within ~1" of the primary. This is, so far, the most stringent limitation to the existence of a third body in the Sirius system.

An additional search for a distant third companion was reported by Bonnet-Bidaud et al. (2000). The motivation for this study were the early reports of a visible companion mentioned above, as well as the proposal put forward by Bonnet-Bidaud & Gry (1991) that a tidal event involving a third body in the Sirius system could have caused the transient reddening reported in antiquity. The authors made ingenious use of their two-epoch coronagraphic imaging of the neighborhood of Sirius, where the image of the bright object was masked off. It allowed the recording of 31 stars in a field with size 4.8×3.6 arcmin, where only the innermost 30 arcsec (where the masked image of Sirius was located) could not be investigated.

The image of the Sirius neighborhood with the bright star masked off, obtained in 1999, was compared with a similar image taken in 1985. The authors searched for objects that shared the proper motion of the Sirius system during these 14 years, in which the projected motion amounted to 17 arcsec. This method identifies possible companions by requiring them

---

[6]Spectral class for stars with an effective temperature lower than class M, with metallic hydrides and neutral alkali metals as the major spectroscopic signatures.

to follow the same projected motion as that of Sirius, mostly being rather distant from Sirius itself. Needless to say, such a companion star was not found.

The combination of the 1985 and 1999 coronographic images of Sirius is shown in Figure 4.4. One image was subtracted from the other, thus the imperfect matching of the point-spread-function of the stellar images caused positive and negative parts of the images. The stars themselves are identified in Figure 4.5.

The conclusion of Bonnet-Bidaud et al. (2000) was that the most likely explanation for the reports on the apparent third body in the Sirius system is the presence, in the 1920s, of one of the background stars seen in projection close to the system. This object (no. 2 in Figure 4.5), an I=12.63 star, was only 6.9 arcsec away from Sirius in 1937. A fainter star (no. 20 in Figure 4.5) with I=14.69 was only ∼2 arcsec away in 1934. Visual observers, therefore, would have seen a faint star close to the α CMa system during these years and could have mistaken it as a member of a triple system.

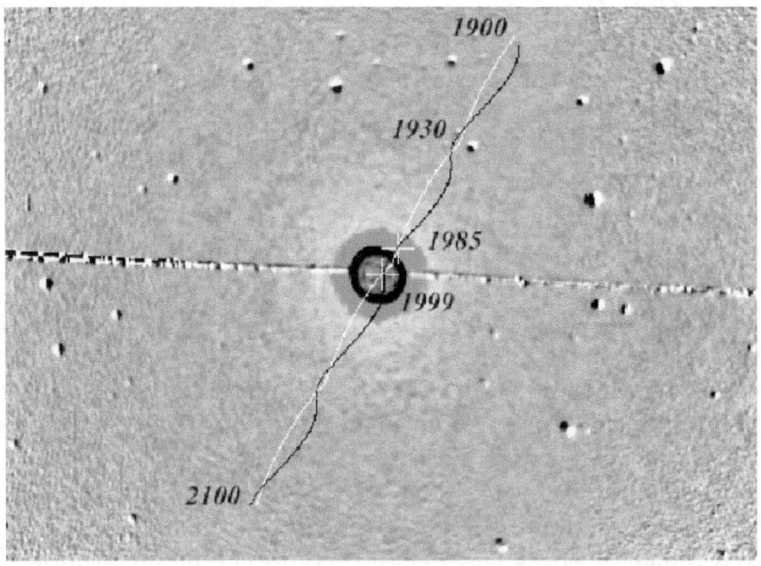

Figure 4.4. Track of Sirius on the background of neighboring stars, from Bonnet-Bidaud et al. (2000), showing stars that could have been mistaken as physical companions in the Sirius system if seen at the proper time. Sirius is blocked off in the image for the epoch of observation, 2000, and its track as well as that for Sirius B are plotted.

This, indeed, could be the conclusion of the controversy on the possible existence of a third body in the Sirius system, if not for the puzzle presented by the lack of claims about a companion during the closing years of

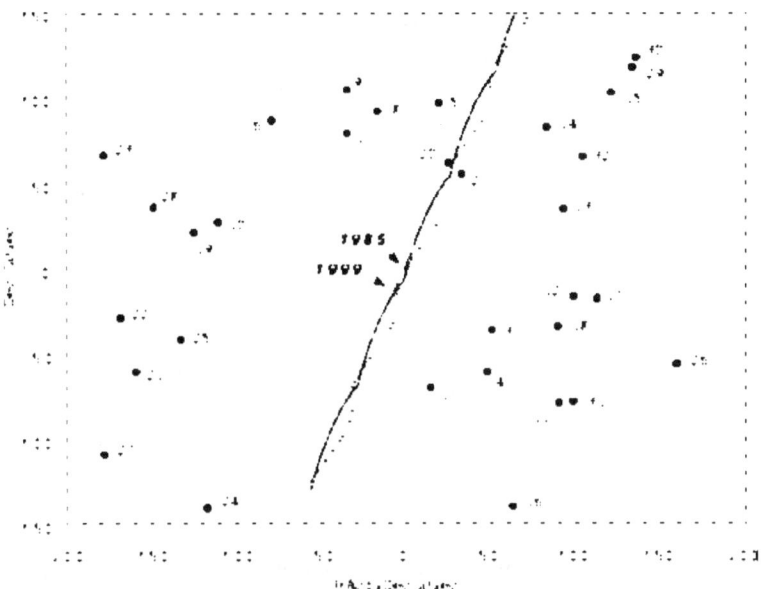

Figure 4.5.   Stellar identifications on the image of the Sirius neighborhood from Bonnet-Bidaud et al. (2000). The axes' units are arcsec.

the nineteenth century. Using the data presented by Bonnet-Bidaud et al. (2000), it is clear that in the last decade of the nineteenth century Sirius and its known companion passed close to star no. 5, a 14 mag object, that should have been seen projected at somewhat larger separation than that between the A and B components (see Figure 4.4). The last years of the nineteenth century saw some intense observations of the α CMa system in order to determine the binary orbit. However, the astronomical literature in that century does not contain reports of a third star associated with Sirius. Some visual observers reported that the supposed C companion was a bright star, with V≈2.5–3 (van den Bos 1929; Finsen 1929) and not m_v=12 as mentioned by Bonnet-Bidaud et al. If this would have been their star no. 2, which seems to be their best candidate, it would have to be an extremely variable one to have dimmed by ∼8 mag in ∼70 years.

The lack of a detection by Bonnet-Bidaud et al. (2000) does not completely rule out the existence of a third companion in the Sirius system. First, the limiting magnitude for the detection of stars with their coronagraphic observations was only V≃17 mag and I≃16, thus there is some leeway for a fainter companion such as a brown dwarf (Kelu-1 at the distance of Sirius would be an m_v ≃18–19 mag star, using the absolute magnitude of Ruiz et al. 1997). Second, the innermost 30 arcsec close to Sirius could not be surveyed at all by Bonnet-Bidaud et al., since this region was masked off in the coronagraph. Neither was a sensitive proper motion study performed with the HST for companions fainter than could have been detected by Bonnet-Bidaud et al. Therefore, a close companion, especially at a suitable orbital inclination, could still exist in the Sirius system. Bonnet-Bidaud et al. estimate the probability of such a companion, in an orbit with a=230 a.u. and e=0.9, as about 1%.

The recent searches for a third companion using the HST seem to rule out most of the phase space where a third companion might hide. The regions yet to be explored are the very close orbits around Sirius A or Sirius B, within about 1 arcsec of each. For a significant improvement in detection capability, one would have to wait until the Space Interferometry Mission or a similar ultra-high angular resolution imaging instrument, with high rejection capability of the light from Sirius A, would become operational.

4.4. Conclusions

The theoretical prediction by Bessel of a second body modifying the space trajectory of Sirius was proven correct with the discovery of Sirius B by Alvan Clark Jr. and with the follow-up observations. This, together with a good determination of the parallax of Sirius, allowed the calculation of the masses of the two components. The puzzle remained, because in the nineteenth century and in the first decades of the $20th$ there was no good explanation for the nature of a body with the mass of Sirius B but of this extreme faintness.

A further puzzle regarding the Sirius system, of reports of a third body in the system, seems to be almost completely solved since such a body was not observed by modern astronomers. The parameter space still allowing the existence of such a component has been reduced very significantly, but has not been completely closed. Sirius might indeed harbor planets, but whether these were seen by astronomers in the first half of the twentieth century seems doubtful.

Chapter 5
Modern optical measurements

The Sirius system was measured many times in the twentieth century, using different observational methods and in a variety of spectral bands. Some of these measurements were mentioned previously. I shall describe below some of the many measurements and methods used. Before continuing, though, I describe first the various nomenclatures used by modern astronomers to indicate Sirius itself. This is of interest when searching for information about Sirius in existing catalogs and publications.

One star-naming method was already described before. This is the giving of names by brightness among the stars of the constellations. The method is valid as long as the borders of the constellation are well-understood and accepted by the community. As mentioned above, this method was invented by Johann Bayer in 1603 and uses Greek letters to indicate relative brightness. Bayer called the brightest star of a constellation α, the second β, and so forth. However, the borders of the constellations themselves were kept fluid for a long time, and until 1930 celestial cartographers selected their own constellation boundaries.

The first time borders of constellations were drawn on an astronomical map was in 1801, by Bode[1] (Norton 1969). Before, stars would be plotted sometimes on celestial maps, but most often not (see e.g., Figures 2.15 and 5.1). The constellations were represented by stylized figures. Stars received Greek letters, according to their brightness in their constellation. Stars fainter than the limit set by the number of letters in the Greek alphabet received numbers in the Flamsteed[2] catalog of 1725 (Norton 1969). The Flamsteed numbers go by order of the Right Ascension of a star within the constellation.

The boundaries of constellations, as drawn on different star maps and atlases, changed according to the whim of the plotter, as Figure 5.2 shows. Only in 1930 did the International Astronomical Union adopt a modified

---

[1] Johan Elert Bode (1747–1826), astronomer who worked in Berlin. Bode produced a stellar atlas containing some 17,000 stars and plotted in it almost all the constellations then invented (*Uranometria*, published in 1801). In 1782 he published a German version of the Flamsteed stellar atlas.

[2] John Flamsteed (1646–1719), British astronomer, appointed Astronomer Royal in 1675.

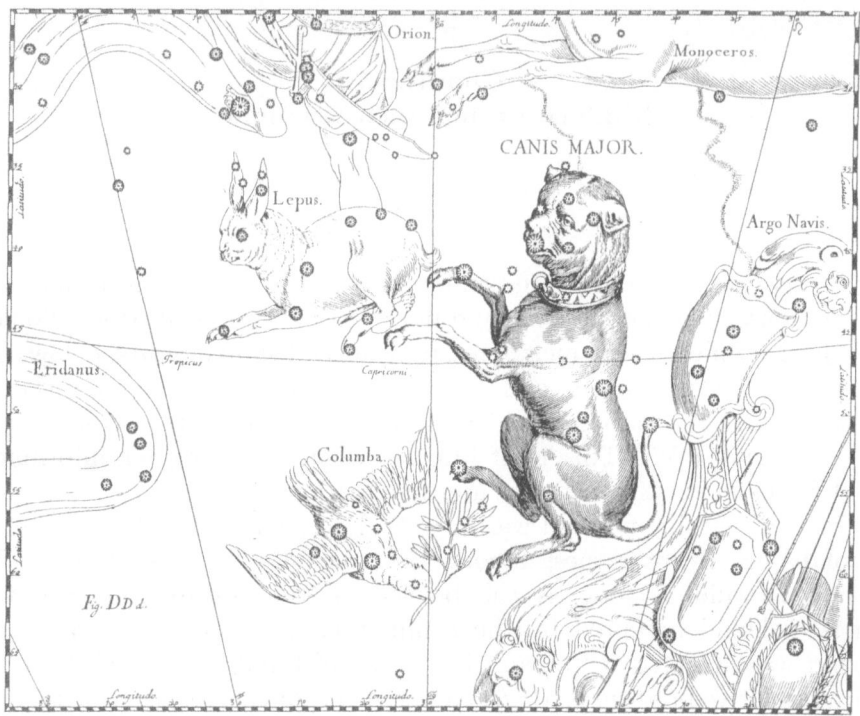

Figure 5.1. Section of a star map showing CMa, from the Star Atlas drawn by Johannes Hevelius in the sixteenth century.

set of borders that followed the arcs of Right Ascension (α) and Declination (δ) for the epoch January 1, 1875. The IAU delineation of the constellation borders survives to this date.

More modern catalogs follow a similar convention of listing stars in order of increasing α, but do away with the borders of constellations. Such catalogs, dating from the nineteenth and twentieth centuries, are the Henry Draper (HD), the Bright Stars (HR), the Gliese catalog of nearby stars, etc. A star catalog compiled by Argelander in the mid-1800s at the Bonn Observatory contained about 300,000 stars arranged in one-degree wide strips of equal declination. It is known as the *Bonner Durchmusterung* and its stars carry the prefix BD with a number for their degree of declination and another number for its order (of increasing α) within the declination strip. The Sloan Digital Sky Survey (SDSS data release 6) imaging catalog contains 287 million objects; the entries are identified as SDSS JHHMMSS. ss+DDMMSS.s using truncated, not rounded, coordinates for epoch 2002.

Sirius is listed in all the stellar catalogs. It is HD 48915, HR 2491, as well as ADS 5423 A and B in the double-star catalog and BD $-16°$ 1591 A

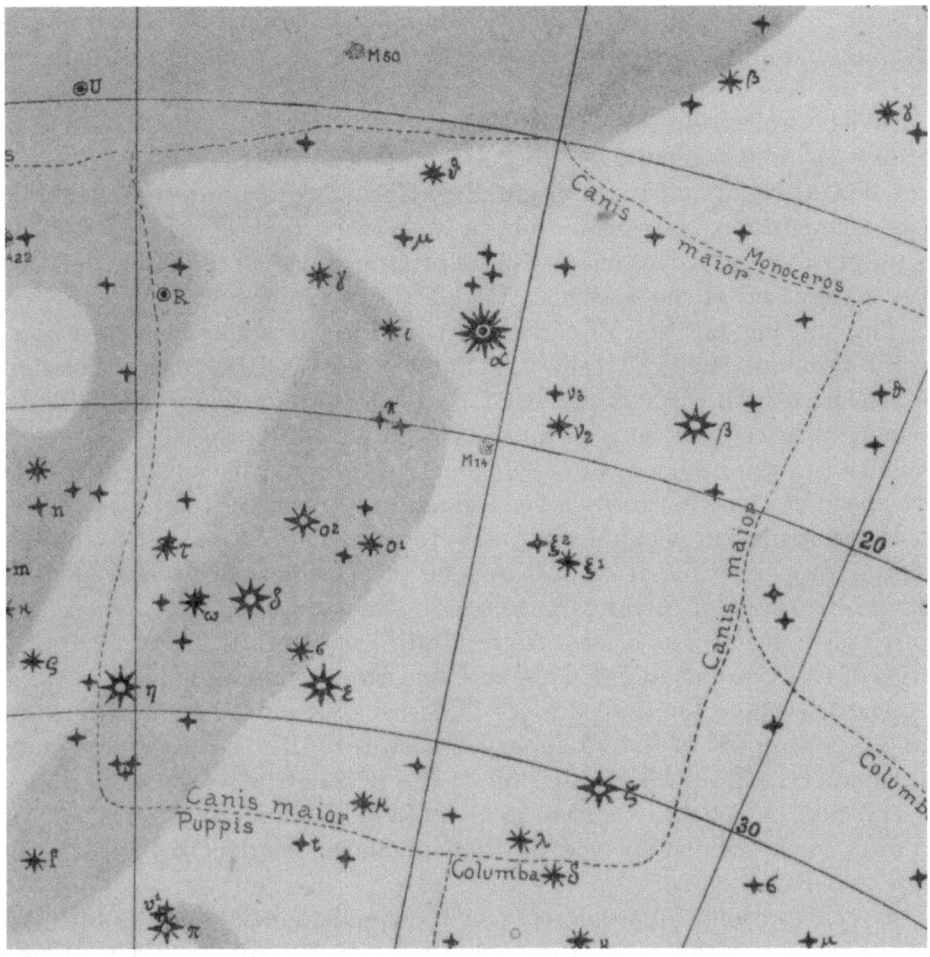

Figure 5.2. The fluid borders of constellations (here of Canis Majoris) were plotted so as late as 1923 (from Becker 1923).

and B in the *Bonner Durchmusterung*. In the Luyten proper motion catalog it is LPM 243 and in the Position and Proper Motion catalog it is PPM 217626. The SIMBAD[3] data base lists 53 different identifiers for Sirius. In order to include Sirius and other stars in these catalogs, their celestial positions must be accurately measured.

5.1. Astrometry

Astrometry is the astronomical discipline where positions of celestial objects are measured with high accuracy. Given that one can measure only angular distances between stars, and than an angle of one arc-second is what an

[3]Stellar data base kept at the Centre de Donnés Astronomiques de Strasbourg.

object 1 cm across subtends when viewed from a distance of 2 km, it is clear
that angular distances much smaller than this angle require tremendous
efforts in order to be measured accurately.

Being a relatively nearby star, Sirius has a large apparent motion on
the sky, as already mentioned above. Gore (1903) and Serviss (1908) noted
that 60,000 years ago it was located on the eastern border of the Milky
Way, whereas it is now located on its western border. This is the result of
its proper motion, a combination of its orbit around the center of the Milky
Way galaxy and of the similar motion of the Sun and Earth.

The shifting position of Sirius on the sky was apparently first noted
by Sir Edmund Halley[4] (Halley 1717). He compared the stellar positions
he measured with those in ancient Greek catalogs and found that the discrepancies in the positions of Sirius, Procyon, and Arcturus could not be
explained as the influence of precession. This implied the existence of relative space motions between these stars and the Solar System.

Parallax measurements for Sirius were reported by Henderson (1839) as
0."25 with a probable error or a quarter of a second of arc; he concluded
that *the parallax of Sirius is not greater than half a second of space, and
that it is probably much less.* Abbe (1867) reported 0".27±0".10 from a
series of observations at the Cape of Good Hope observatory.

Another value for the parallax of Sirius was published by Comstock
(1897) as a report of the Washburn observatory. He used a double-image
micrometer to produce a value of $\pi = +0".31\pm0".03$ for the period of
observation, which lasted from 1893 to 1896. This method relied on the
accurate measurement of the position of Sirius relative to more distant
stars.

A combination of astrometry, which defined well the orbit of Sirius A,
and two-epoch monitoring of its radial velocity by Adams (1902) allowed
the determination of a parallax for Sirius. Adams used a previous value
for the radial velocity, -15.6 km s^{-1} measured in 1890 at Potsdam, and
a new value determined from ten spectra obtained in 1901–1902 which
averaged to -6.87 km s^{-1}, to determine a parallax of 0".21. This was done
by matching the change in radial velocity with that expected, given the
orbital parameters of the binary star.

Van de Kamp & Barcus (1936) used photographs of the Sirius system obtained with the 26-inch McCormick refractor to improve the parallax determination and re-derive the mass ratio of the two stars from the
previously-determined orbit by Volet (1931). A modern depiction of the
orbit was shown in Figure 4.1 in the previous chapter. Its knowledge, in
absolute units, allows the determination of the two stellar masses. Van de
Kamp & Barcus found a mass ratio (mass of the lighter component divided

---

[4]The second Astronomer Royal (1656–1742), well-known for the periodic comet he
identified and is named after him.

by the total mass of the system) of 0.326±0.006, yielding individual masses of 2.6 $M_\odot$ for Sirius A and 1.3 $M_\odot$ for the B companion, for a total system mass of 3.9 $M_\odot$.

A long series of observations of Sirius A and B was collected with the 60-cm refractor of the Bosscha Observatory, at Lembang (Indonesia). These observations consisted of photography with a five-wire grating installed in front of the telescope lens. The distance between these wires was adjustable, to make the distance of the first diffracted image of Sirius A nearly equal to the A–B distance at the time of the observation. Measurements with this setup were reported by Van Albada-Van Dien (1977).

A very comprehensive astrometric study of the Sirius system was by Gatewood & Gatewood (1978). They used 308 photographic plates obtained with the refractors of the Yerkes and Allegheny observatories to re-determine the binary orbit and the parameters of the system. Gatewood & Gatewood determined an even more accurate value for the distance, 2.65±0.03 pc, and for the individual masses of the components: M_1=2.143±0.056 $M_\odot$ and M_2=1.053±0.028 $M_\odot$. This allowed them to reject some of the Hamada & Salpeter (1961) white dwarf models and claim a better fit for white dwarfs with ^{4}He, ^{12}C, and ^{24}Mg compositions. This will be discussed in more detail in Chapter 8.

Jasinta & Hidayat (1999) extended the measurements of van Albada-van Diem (1977) with plates from the same Lembang telescope obtained in 1976-1986. They presented yearly means of the position angle and the distance between the components of the Sirius system. To date, there has been no comprehensive study of Sirius combining all the astrometric material collected by different astronomical observatories (except possibly for unpublished work by J. Holberg).

5.1.1. The *Hipparcos* Satellite

The *Hipparcos* satellite was launched by the European Space Agency (ESA) in 1989 and operated till 1993. The idea of launching and operating an astrometric satellite was discussed within the astronomical community at least since the Prague meeting of the IAU General Assembly in 1967. The name of the mission, which sounds very much like the name of the Greek astronomer Hipparchus, is the acronym for *HIgh Precision PARallax COllecting Satellite*. The task of this mission was the production of the most accurate (to date) catalog of positions for the brighter stars, those with $m_V \leq 9$ (Perryman et al. 1992).

The instrument consisted of a specially designed Schmidt telescope with a 29-cm diameter mirror. The secondary mirror of this telescope served not only to correct the field, but also to project two well-separated sky fields onto the focal plane. This was achieved by cutting the mirror in half and by cementing the halves together at an inclination of 29° to the optical axis of

the telescope; this gave two fields of view each 0°.9 wide separated by 58°.
The spin of the satellite, at 11.25 revolutions per day, carried the fields of
view across the sky. A modulating grid, with lines perpendicular to the scan
direction, was mounted in front of the detector (an Image Dissector Tube)
and a star in either of the fields which entered the detector would give
rise to a modulated signal. This way it was possible to measure with high
accuracy the angular separations of stars along the scan direction. Despite
being left in an highly elliptical transfer orbit, due to the failure of the
apogee boost motor on its way to the intended geo-synchronous location,
Hipparcos succeeded in all its intended tasks.

The *Hipparcos* catalog contains more than 118,000 stars with parallax-
derived distances and accurately measured proper motions. In addition, the
satellite also measured the brightness of the objects it observed. The results
from the *Hipparcos* measurements were combined with cataloged data from
the *Hipparcos* Input Catalog, which contains also radial velocities for all
the *Hipparcos* stars. These were culled from the astronomical literature
and form a very extensive collection of information.

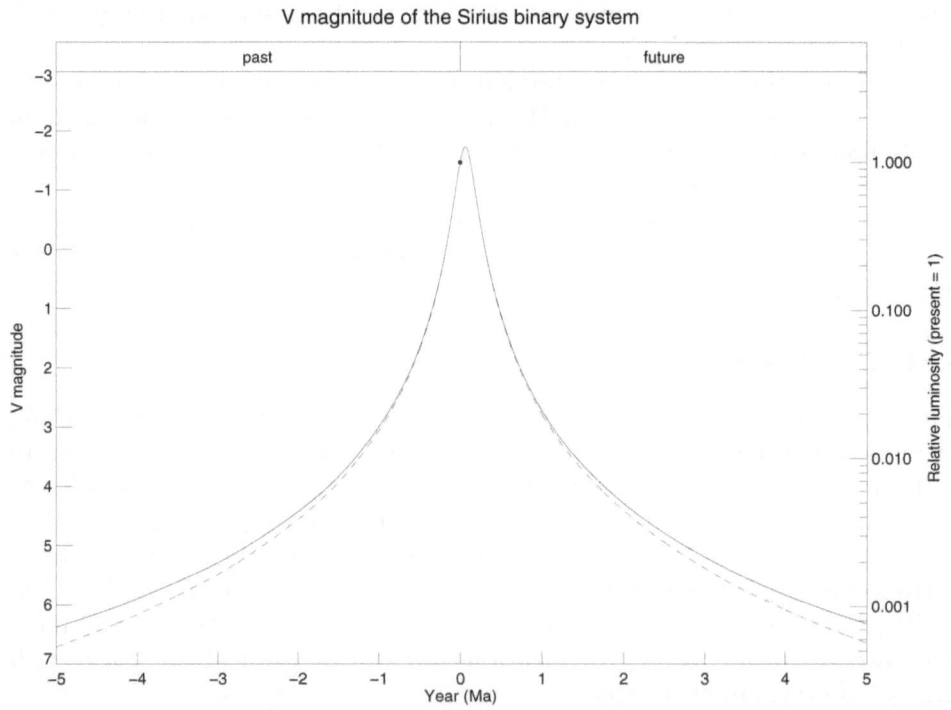

Figure 5.3. The changing distance to Sirius ±5 Myrs around the present time modifies
its apparent magnitude. This plot shows the apparent magnitude of Sirius in absence of
interstellar extinction (*solid line*) and assuming a uniformly-distributed extinction of 3.5
mag kpc^{-1} (*dashed line*). I am indebted to Dr. R. Van Gent for producing this figure.

The combination of various kinds of data, parallax, proper motion, and radial velocity, can be used to calculate the future apparent behavior of Sirius among other apparently bright stars. Tomkin (1998) demonstrated that the relative motions of the Sun and of Sirius are bringing these two bodies closer. From the present-day 2.63 pc distance, Sirius will approach to only 2.39 pc some 50,000 years from now. During this time neither the Sun nor Sirius will evolve significantly, thus their absolute magnitudes will remain approximately the same. For an Earth-bound viewer, Sirius will brighten slightly to look like a −1.64 mag star (instead of the −1.44 mag object it is today) (see Figures 5.3 and 5.4). Sirius will remain the brightest star in our skies for the next 150,000 years; then Vega will surpass it in apparent brightness for the next 250,000 years.

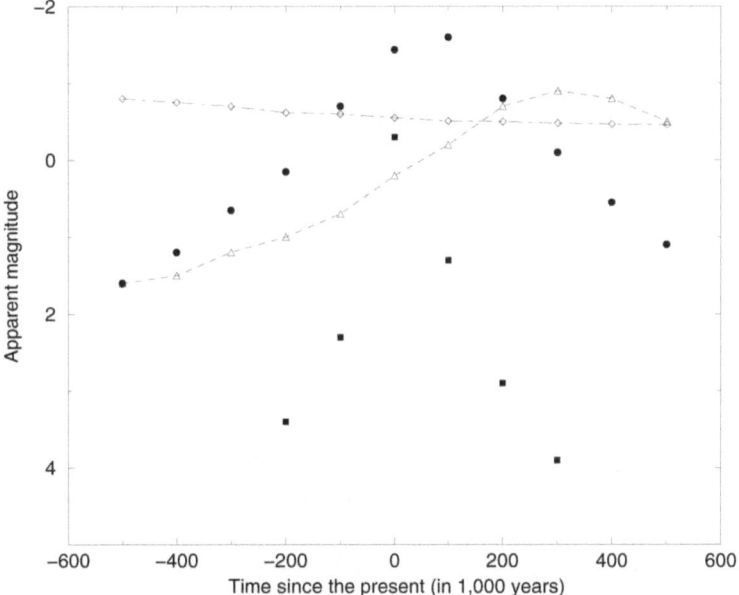

Figure 5.4. The changing distance to Sirius ±0.5 Myrs around the present time also modifies its apparent magnitude in comparison with other bright stars. This plot is following Tomkin (1998). Sirius is represented here by filled circles, α Cen by filled squares, Canopus by open diamonds linked with a dot-dashed line, and Vega by open triangles linked with a dashed line. Sirius is our brightest star only for a limited period.

Note that plots such as that by Tomkin, do not normally account for the effect of the intervening ISM extinction. This might change as a star travels through the Milky Way, given the size of dust complexes (tens of parcsec) and the typical distance travelled by a star ($\sim$10 pc/Myr at 10 km s $^{-1}$). The effect is slight, as Figure 5.3 shows, for the case of a uniform dust distribution. A discussion of the distribution of local interstellar matter will be presented in Chapter 7.1.

5.2. Photometry

Photometry measures accurately the energy output of an astronomical object received by the measuring instrument. This can be a "bolometric" measurement, where the entire spectrum of the object is covered by the instrument, or it can be band photometry. In the latter method a restricted spectral region is separated by means of a filter or filter combinations, which define the center of the band and its width when combined with the response of the detector, and the light output of the source is measured through this band.

A first report of astronomical photometry, measuring bright stars by using "transmission screens" in front of a vacuum thermopile, was by Coblentz (1922). The transmission screens used were essentially wide-band filters, each selecting a certain region of the spectrum to be transmitted to the detector. The filters used in this study were yellow or red glasses, water, or a thick quartz plate. The spectral regions selected with combinations of these filters were 0.3–0.43 μm, 0.43–0.6 μm, 0.6–1.4 μm, and 1.4–4 μm. The results showed that B or A stars have their maximum intensity in the ultraviolet, while cooler stars of types K and M have their radiation peak at 0.7–0.9 μm. Coblentz concluded that the red M stars have an approximate temperature of 3,000 K while the blue B stars have a temperature of 10,000 K. Sirius itself is noted in his Table 1 as a binary of spectral type A, with the brighter component at $m_{Harvard}$=−1.58 and a secondary of 8.5 mag. Presumably the magnitude of Sirius B could not have been measured with this technique and its derivation was through the brightness ratios found by photography or by eye observations. Coblentz concluded that Sirius B must be of spectral type K, but in 1927 he retracted this claim after he found an experimental error in his measurements (Coblentz 1927).

The spectral energy distribution (SED) of Sirius was measured by Abbot (1924) using a radiometer at the Coudé focus of the 100-inch Mt. Wilson telescope that sampled the spectrum dispersed by a prism. The results, from 4940 Å to 2.224 μm and corrected for atmospheric extinction, were fitted to a black-body SED yielding a stellar temperature of 11,000 K and a stellar radius twice to three times that of the Sun. The work relies on a comparison with measurements of the Sun made with the same instrument after being attenuated by a factor of 0.89×10^{10}; Abbot remarked on the difficulty of performing such a comparison with the data then available, pointing out the fact that:

> the sun's energy curve fits no black-body curve whatever. It is too high in the infra-red and too low in the violet.

The discrepancy was attributed by Abbot to the different wavelengths of the observations probing different depths, thus different temperatures, in the solar interior.

Huffer & Whitford (1934) described photoelectric measurements of Sirius B performed at the 100-inch Mt. Wilson reflector when Sirius B was furthest away from A. They used small apertures for the measurements and carefully attempted to subtract the scattered light contribution from Sirius A, to obtain a visual magnitude of 7.3 for the WD.

Stebbins (1950) described the early times of photoelectric photometry, which included searching for better cathode materials, cooling detectors to dry-ice temperatures, and obviously operating at larger telescopes. Sirius was a target for his studies, in an attempt to determine its true and accurate visual magnitude. This was determined through a comparison with stars in the North Polar Sequence[5], or bright stars in the vicinity of Sirius, and yielded m_V(Sirius)=-1.42 ± 0.03, about 0.16 mag fainter that the visual magnitude adopted till then. The measurement was done, by the way, with the 36-inch Crossley reflector at the Lick Observatory, and by using a "wire screen" (presumably a neutral-density filter) that attenuated Sirius by five magnitudes. It refers to a spectral band with an effective wavelength of 5380 Å.

Eggen (1950) described photoelectric measurements of 180 nearby stars, among which he observed Sirius. His results are given in the photometric system used in the 1950s, photographic and visual magnitudes and color indices. The observations of Sirius, a very bright object, were done with the Lick Observatory 12-inch refractor equipped with a screen reducing the brightness by 4.82 mag. The values obtained for Sirius are P_{g_p}=-1.53 mag (total photographic magnitude) and a color index between the photographic and visual bands of C_p=-0.07.

Eggen & Greenstein (1965) listed spectroscopic and photometric data for a sample of 166 white dwarfs, one of which was Sirius B. They give for this white dwarf the following photometric measurements: V=8.3, B–V=-0.12, and U–B=-1.03, with the colors estimated from high-dispersion spectra and the profile of the Hγ line, based on observations by Greenstein & Oke.

Parenthetically, note that if one knows the B–V color of a star this allows the determination of its effective temperature T_{eff}. This is because the main-sequence stars follow a definite relation between these two parameters, as shown first by Johnson (1966). The hotter stars on the main-sequence were shown to follow such a relation by Code et al. (1976). A

---

[5] A set of stars in the vicinity of the North Celestial Pole whose positions and apparent magnitudes were determined with high precision in order that they could serve as standards for stellar photometry.

re-determination of this relation was done by Sekiguchi & Fukugita (2000). The accuracy is 0.015–0.02 mag for stars from F0 to K0, and is reduced somewhat for stars as late as K5.

An interesting and original attempt to measure the magnitude and colors of Sirius B was by Rakos & Havlen (1977) using the ESO 1-m telescope on January 28 and 29, 1975, at almost maximal separation between the two components of the system. They used an "area scanner" photometer, which is essentially a scanning slit passing back and forth in the telescope focal plane and feeding a regular photometer. In the pre-CCD era, the area scanner could produce a kind of one-dimensional image, which was sampled with a linear, photon-counting detector.

Rakos & Havlen (1977) used a 1"×9" slit for the Sirius observations and scanned an area 13" long, from Sirius B in the direction of Sirius A. This was so that they could account properly for the scattered light of Sirius A when performing photometry of the B component. They found the following averaged values for Sirius B: V=8.44±0.03; B–V=–0.03±0.03; U–B=–1.04±0.03, b–y=–0.04±0.01; m_1=0.17±0.01; c_1=–0.21±0.02; Hβ= 2.77±0.03 and u–b=+0.05±0.03. The photometric values measured by Rakos & Havlen, in the broad-band Johnson colors and in the intermediate-band Strömgren filters (they are represented by the lower case u, b , v, and y filter names), were in agreement with the values obtained for other DA white dwarfs. They allowed the authors to determine an effective temperature for Sirius B of 32,000±1,000 K. This T_{eff} could not support a young age for the white dwarf, as Rakos (1974) earlier proposed based on similar observations.

The photometric measurement of Sirius itself is a difficult task, the star being so bright that one needs to calibrate well the means of reducing the brightness to measurable values with a well-calibrated photometer. In this context, the remark by Johnson (1980) that Sirius became fainter by 0.05 mag since his measurements in the 1950s is relevant. One wonders whether such a small, secular change may not be the result of a small change in calibration.

Another use of the *Hipparcos* data, which was mentioned above, was a search for possible small-amplitude photometric variability of Sirius (Percy & Wilson 2000). The trigger for such a search was a proposal by Struve (1955) that there exists a group of variable stars of types B7 to A3 near the main-sequence whose light changes with small amplitudes and with periods of a few hours. Photometric variability, due to non-radial pulsations, has been found in slowly pulsating mid-B stars and in mid-F types (γ Doradus type), thus one could expect Sirius to belong to this class. Note the prediction of Kervella et al. (2003) that, if Sirius A has radial oscillations, the frequencies of these modes would be separated by 81–82 μHz.

I am not aware of any astero-seismology experiment involving Sirius that could have validated this prediction.

Percy & Wilson (2000) searched the individual observations of Sirius by the Hipparcos satellite for such a variability. The *Hipparcos* data set[6] includes an epoch photometry catalog of all observed objects. This contains brightness measurements every 20 min or so for several hours, and sometimes for a day or more, with gaps of 20–30 days. The photometry is in the peculiar *Hipparcos* system, a broad band covering a spectral range wider than the Johnson B and V bands. The result of an auto-correlation analysis of all *Hipparcos* measurements of Sirius shows a possible periodicity of four days, but the level of the variation, a few 0.01 mag, is similar to the baseline scatter in the auto-correlation. According to Percy & Wilson, this indicates that the variability is instrumental and is due to the exceptional brightness of Sirius.

Adelman et al. (2000) included Sirius among the A0–A2 stars in which they searched for variability using the Hipparcos photometry. They reported a formal detection for Sirius, with an amplitude of 0.19 mag., but also mentioned that this detection is doubtful (no cause given).

5.3. Spectroscopy

Fraunhofer[7] observed the spectrum of Sirius along with that of other bright stars in the first half of the nineteenth century. This was just after he described some 324 absorption lines in the spectrum of the Sun. While observing Sirius, he managed to detect only three wide and dark lines, one in the green and two in the blue part of the spectrum (Abetti 1954). These features were present also in the spectrum of Castor. They are the prominent Balmer lines, as Figures 5.5 and 5.6 show.

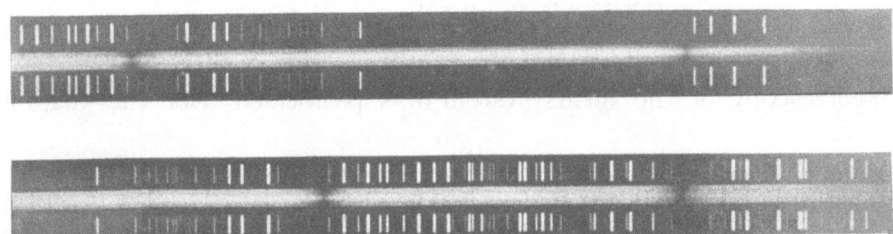

Figure 5.5. The optical spectrum of Sirius A shown here extends from 4,000 Å to 4,900 Å. Its dominant features are the deep and broad Balmer (Hydrogen) lines, characteristic of a main-sequence A-type star.

---

[6]Available electronically at http://astro.estec.esa.nl/Hipparcos/hipparcos.html

[7]Joseph von Fraunhofer (1787–1826), German optician and astronomer. Discovered absorption lines in the Solar spectrum in 1815.

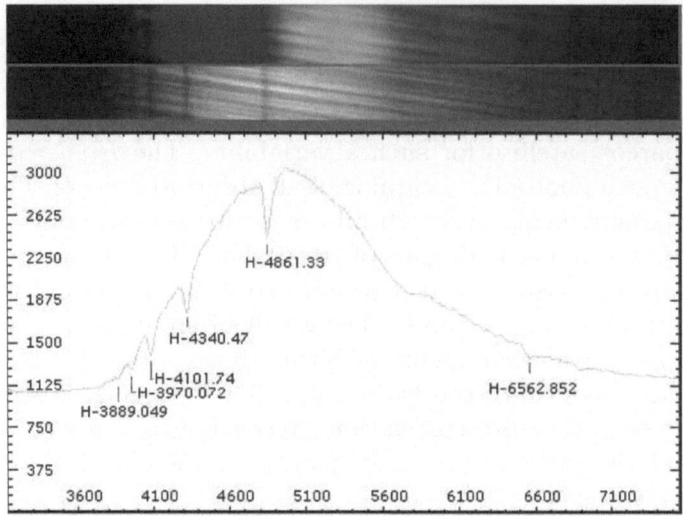

Figure 5.6. The optical spectrum of Sirius A (image shown with two different display scales at the top and spectrum tracing shown at the bottom) shows the deep and broad Balmer (Hydrogen) lines, characteristic of a main-sequence A-type star. The spectrum was obtained with a small telescope and a transmission grating, and is included here courtesy of John A. Blackwell, Northwood Ridge Observatory.

The Fraunhofer lines in the spectrum of Sirius were studied by Huggins[8] who found that the lines were slightly shifted, in comparison with laboratory spectra. He showed, therefore, that a Doppler effect was present, allowing the measurement of the velocity of Sirius in the laboratory frame of reference at about 30 miles/sec (Abetti 1954).

A simplified description of the spectrum of Sirius, as understood at the beginning of the twentieth century, was presented by Serviss (1908):

> *The spectrum of Sirius is typical of a class known as the Sirian stars, which include, perhaps, half of all that are visible, and which are characterized by a brilliant white color and broad absorption bands, indicating the presence of enormous quantities of hydrogen.*

Spectroscopy of the Sirius system was performed over the years in order to:

1. determine the exact spectral type of the A component,
2. establish the effective temperature of the B component,
3. measure the star's magnetic field,
4. measure the line-of-sight velocity of the system,
5. measure the rotation of Sirius A,
6. derive the gravitational redshift of the B component,
7. determine the properties of the interstellar medium between us and the Sirius system.

[8]William Huggins (1824–1910), British astronomer and investigator of stellar spectra.

One of the first reported observations was by Vogel (1895), discussing the Helium lines observed in Sirius. The report should be of interest to students of the history of science, because Vogel mentions the identification of a strange spectral line, seen in the Sun and and close in wavelength to the Sodium doublet, as produced by a gas contained in the "rare mineral *Clèveite*[9]". Vogel identified the same substance in the spectrum of β Lyrae, as well as detecting the presence of two orbiting bodies in this system based on the motion of the lines (Doppler shift).

Vogel gave a detailed account of his spectral investigations "of all stars of the first class down to the fifth magnitude" and classified α CMa as a star belonging to spectral class Ia2, with

> *spectra in which lines of other metals (calcium, magnesium, sodium) appear, in addition to the hydrogen lines, but which contain no lines of Clèveite gas. The calcium line at $\lambda 3934$ Å in these spectra appears sharply defined; its breadth is not nearly equal to that of the hydrogen lines. The spectral lines of other metals are delicate, and are not easily recognized with low dispersion.*

The lines of the mysterious *Clèveite* gas were recognized in 1895 by William Ramsey[10] as belonging to the element Helium.

Accurate spectroscopic measurements of Sirius allowed Campbell (1905) to disentangle its systemic velocity from the orbital motion. The solution used the orbital elements determined by Zwiers (1899), a determination of the parallax from the Cape Observatory (Gill 1898), and a value for the mass ratio between the A and B components: 2.20:1.04 from Auwers (1892, cf. Campbell 1905). Campbell determined a space velocity of –7.4 km s^{-1} by subtracting the predicted orbital motion from the measured radial velocity.

Adams (1911) reported on line shifting in the spectrum of Sirius that could be attributed to pressure effects. He measured the stellar spectral lines along with reference emission lines from an iron arc and found a systematic positive displacement, the stellar lines being redder than those of the arc lamp by 0.014 Å=0.90±0.13 km [presumably km s^{-1}]. Adams mentioned that similar results, of systematic redshifts (longer wavelengths), were obtained for lines observed on the solar limb. He concluded that a comparison with Sirius indicated a pressure higher by 12 atmospheres in Sirius than in the similar layer of the Sun.

---

[9]Clèveite is a mineral named after the Swedish chemist Per Theodor Cleve (1840–1905). It is an oxide of Uranium and Lead, which releases Helium when heated with acid.

[10]Sir William Ramsey (1852–1916), Scottish chemist. Received the 1904 Nobel Prize for Chemistry.

Lindblad (1922) described a "spectrophotometric method for determining absolute magnitudes". The method was based on the relative intensities of the stellar continuum on both sides of the wavelength $\lambda 3907$ Å, and on the relative intensities of the CN bands relative to the adjacent continuum. Lindblad compared Sirius with the star o_2 Eridani, then known as an A0 star of an absolute magnitude $+10.3$, and concluded that the hydrogen lines in the latter were strongly widened while the spectra showed a faint extension of the continuum into the ultraviolet. This went on much farther to short wavelengths than seen in spectra of "ordinary" A stars. Lindblad essentially identified the spectral characteristics of a white dwarf, without actually recognizing the star as such.

It is important at this point to also discuss briefly the development of stellar classification techniques. Huggins (1877) described early attempts to photograph the spectra of stars. Bright objects, such as Sirius and Vega, required 15–20 min of exposure with an 18-inch reflector and a prism spectrometer to record their spectra on "dry process" photographic plates. Based on these spectra, he published later an atlas of representative spectra, as an attempt to systematize the information spectra of stars contained.

Angelo Secchi, the director of the Collegio Romano observatory, described a multi-prism spectrometer used to characterize visually the spectra of the bright stars (Secchi 1863). He wrote about Sirius

> *Les étoiles blanches que j'ai examinées présentent surtout des interruptions notables ...le spectre de Sirius et de Rigel qui ont une large bande à la limite du bleu et du vert, avec deux autres, dont l'une dans le bleu et l'autre dans le violet.*
>
> (The white stars I examined show notable interruptions...the spectrum of Sirius and of Rigel have a large band at the limit of the blue and the green, with two others, one in the blue and the other in the violet).

These features are the Balmer lines, with the most prominent being Hβ. Secchi mentioned that these deep absorption lines were already recognized by Fraunhofer.

Later, Secchi (1868) classified some red stars, but also mentioned Sirius as belonging to the 1st type, together with α Lyr [modern type A0V], while the prototypes for the 2nd type were Pollux [modern type K0 IIIb] and the Sun, for the 3rd type were α Ori [modern type M2Iab] and α Her [modern type M5IIvar/(G5III+F2V)], etc. Secchi essentially attempted a very crude stellar classification based on only four separate types.

Pickering (1890) undertook a titanic project of massive stellar spectroscopic classification. He and the staff at the Harvard College Observatory used a 20-cm diameter lens with a focal length of 115-cm preceded by a thin prism to project the spectra of all the stars in fields $10° \times 10°$ wide on large (20×25 cm) glass plates. The prism produced long spectra and these were trailed perpendicular to the dispersion during the 5 min exposures. With this setup, Pickering's assistant, Antonina Mury, produced a catalog of 10351 stars north of declination −25° known as the Draper Catalog.

The spectral classification of stars was extended by Annie Jump Cannon, who classified 1122 stars whose spectra were photographed in Peru with a 33-cm telescope equipped with objective prisms. This work was published as Cannon & Pickering (1901) and introduced the now-accepted spectral classification A, F, G, K and M types (with the now deleted type N, and with types O and B added). The final Henry Draper catalog, containing the spectral classification of 225300 stars, was published as dedicated issues of the Annals of the Harvard Observatory between the years 1918 and 1925, all contributions being authored by Cannon & Pickering. Sirius appears in Volume 92 of the Annals, on p. 248, with the entry H.D. 48915, an A0 star with a photographic magnitude of −1.58.

The spectral classification and the presence or absence of certain features indicate a basic property of a star, its effective temperature. With its A0 classification, the temperature of Sirius A was determined to be approximately 10,000 K.

A special method of characterizing the properties of a star based on relative gradients, i.e., spectral slopes (or colors in logarithmic units), was used by Gascoigne (1950). Among the 166 southern stars he observed, α CMa stands out with a much bluer color than measured by photometry or than its spectral type would warrant. A thorough comparison with other measurements failed to reveal a cause for the discrepancy, which remained unexplained.

The abundance of Oxygen relative to Hydrogen in the envelope of Sirius was measured by Keenan & Hynek (1950) through the estimation of the intensity of the near-infrared OI triplet of lines at 7771.96, 7774.18, and 7775.40 Å. They found an abundance ratio OI/HI (by number) of 2×10^{-3}.

In the 1960s several spectroscopic analyses of Sirius were carried out in the photographic spectral region. The first, by Kohl (1964), revealed that Sirius showed the abundance characteristics of an Am star when a detailed model atmosphere analysis was done, but that it would be classified as a normal A star on the MK system[11]. Subsequently, abundance analyses were performed by Strom et al. (1966), Latham (1969), and Gehlich (1969). Strom et al. were the first to classify Sirius as an A-type metallic-lined star (Am). The issue of Am stars is discussed further below.

An attempt to obtain a high-dispersion spectrum of Sirius B was reported by Kodaira (1967). This was done with the Coudé spectrograph on the 74-inch Okayama reflector. The spectrum was traced with a

---

[11]The MK system is a way of classifying stars from their spectra introduced in the mid-1940s by Morgan & Keenan. It subdivides the stars according to spectral types and luminosity classes.

microdensitometer and the light from Sirius A that was scattered into the Sirius B spectrum was subtracted off. Kodaira measured the equivalent widths of the Balmer lines, mainly Hγ, and compared these to model predictions to obtain a WD temperature in the range of 14,800–17,000 K.

Using the Meudon Solar Tower spectrograph at a resolution of $(35–68) \times 10^{-3}$ Å Freire et al. (1977) claimed to have observed an asymmetry of the Ca II K line in the spectrum of Sirius. They interpreted this as a signature of velocity fields in the stellar atmosphere, perhaps triggered by the companion star. This, however, was disputed by Griffin & Griffin (1979) using a high-resolution spectrum from the Mt. Wilson Coudé spectrograph, and by Czarny & Felenbok (1979) with a repeat observation using the same Meudon instrument, which failed to reproduce the reported line asymmetry.

The abundance of Boron in Sirius was studied by Praderie et al. (1977), in the context of an effort to understand the fate of cosmologically-relevant species. They used for this purpose the high-resolution UV spectrum obtained by *Copernicus* and measured a rotational velocity $v \sin i$=10±2 km s^{-1}. Praderie et al. found Boron to be under-abundant in Sirius, with only an upper limit for [B/H]$< 5 \times 10^{-12}$ (by number). This is a deficiency factor of at least 20× relative to the cosmic abundance[12], which they interpreted as the signature of either an interaction with the companion star or of diffusion of elements in the atmosphere of Sirius A.

Lambert et al. (1982) used the McDonald 2.7-m reflector at resolutions of 0.2 or 0.3 Å to study the abundances of Carbon, Nitrogen, and Oxygen in Sirius. They measured almost all the detectable C I, N I, and O I lines in the spectral interval 3500–11000 Å, with the conclusion that Sirius is deficient in Carbon relative to Vega by $10^{0.6}$, is over-abundant in N by $10^{0.22}$, and is deficient in Oxygen by $10^{0.27}$. One of the possibilities they tested, concerning these specific abundances, is that Sirius A acquired some CNO-processed material from the progenitor star of the now-WD Sirius B. They mentioned that the sum of the CNO abundances must be conserved by the CNO cycle, because this cycle fuses Hydrogen to Helium while using the C, N, and O nuclei as catalyzers. As this sum is $10^{9.15}$ for the Sun while it is $10^{8.76}$ for Sirius, this would indicate that Sirius is very metal-poor, with [Fe/H]$\approx$–0.8, which seems improbable. The abundance of elements heavier than Helium is expressed through the total abundance of "metals", marked as Z, but sometimes it can be expressed as just the abundance of Iron, using its ratio to Hydrogen:

$$[Fe/H] = \log[Fe/H]_{star} - \log[Fe/H]_{\odot} \tag{5.1}$$

[12]The cosmic abundance of elements is an average of elemental abundances as measured in various celestial sources, from galaxies and intergalactic medium to meteorites from within the Solar System.

Note that from a stellar evolutionary point of view there is no direct connection between the CNO abundance and the Iron abundance. Because of the low total CNO abundance, Lambert et al. discarded the hypothesis of a CNO-processed mass transfer as a possible explanation for the abundance anomalies encountered in Sirius.

Sadakane & Ueta (1989) analyzed data from the high-resolution spectral atlas of Sirius produced by Kurucz & Furenlid (1979). They found that while five "metals" (Mg, Al, Si, Mn, and Ni) have solar abundances to within ±0.3 dex[13], five other elements (V, Cr, Sr, Y, and Zr) are over-abundant relative to the Sun. The abundance plot shown in their paper as Figure 2 indicates that elements heavier than Ti seem to be over-abundant, confirming the classification of Sirius as a metallic-line star (Am). On the other hand, both Nitrogen and Sulfur seem to have solar abundances (Sadakane & Okyudo 1989). The statement about the Nitrogen abundance being approximately solar is in conflict with the statement about its over-abundance claimed by Lambert et al. (1982).

Cayrel de Strobel et al. (1992) included Sirius among the stars for which they checked the Iron abundance. They listed 12 values from the literature for [Fe/H] that averaged to 0.56±0.27 (with a median value of +0.7), implying that Sirius is overabundant in Iron by a factor of 5.

Hill & Landstreet (1993) determined abundances in the atmosphere of Sirius along with five other narrow-lines A-type stars. Their analysis was based on spectra obtained at the Coudé focus of the 1.2-m Dominion Astrophysical Observatory telescope, with a resolving power $R = \lambda/\Delta\lambda \approx 50,000$. Hill & Landstreet confirmed the already-known fact that Sirius is over-abundant in Iron and under-abundant in Calcium, as are other Am stars. The Chromium abundance was studied by Ziznovsky & Zverko (1995), who found that Cr in Sirius has a normal abundance.

Holweger et al. (1999) used high-resolution spectroscopy to search for narrow absorption lines in the center of the photospheric Ca II K and Na D lines in A stars. The goal was to determine whether Ca K or Na shells exist in A stars. Sirius was one of their sample objects in which such cores were not detected, with an upper limit in equivalent width of $\sim$1 mÅ.

A measurement of spectral lines in the 3900–9300 Å region was published by Zhao et al. (2000). The compilation was based on high signal-to-noise observations of Sirius obtained with the 2.16-m telescope of the Beijing Astronomical Observatory and with a Coudé echelle spectrograph. The resolving power used was $\sim$40,000 and the authors tabulated 546 lines with their identification and equivalent widths. The specific improvement of this data set compared with previous high-resolution studies of Sirius (e.g.,

---

[13]The exponent of the power of ten that most closely approximates a number. Thus 0.3 dex represents a factor of 1.995.

Strom et al. 1966; Sadakane & Ueta 1989), lies in the higher accuracy of measurement of the fainter spectral lines, i.e., those with equivalent widths smaller than 25 mÅ.

The abundance study based on these values was published by Qiu et al. (2001). The authors adopted an effective temperature of T_{eff}=9880, a surface gravity log g=4.40, and a micro-turbulence parameter ξ=1.50 km s^{-1} with which they selected model stellar atmospheres. The resultant abundances were found to be anomalous relative to the Sun, with Sirius being metal-rich by [Fe/H]=0.50 dex ($\sim$ 3.16×) relative to the Sun, and overabundant by a full order of magnitude relative to Vega. In particular, and in comparison with the study of Lambert et al. (1982), Qiu et al. found C and O to be underabundant, while the abundance of N was approximately Solar. Qiu et al. also measured the radial velocity of Sirius A to be 13.49± 0.26 km s^{-1} (epochs February 1997 and April 1999).

A comparison of measured abundances for Sirius with values predicted by stellar evolutionary models was presented by Richier et al. (2000) in an attempt to understand the Am and Fm stars. Their basic assumption was that turbulence is competing with atomic diffusion in AmFm stars, thus the surface abundances (for stars of the same mass and initial composition) depend only on the depth of the zone mixed by turbulence. Their Figure 18 shows a comparison of model and measured abundances for Sirius and demonstrates the overabundance of the Silicon-Phosphorus-Sulfur element group, as well as that of the Iron group elements.

The conclusion from the comparison of observations with the theory, for the case of Sirius and given the peculiar mix of elements measured by Richier et al. (2000), was that the mixing depth[14] resides probably at $10^{-5.3}$–$10^{-4.6}$ $M_{\odot}$ below the stellar surface[15] at an age of 100 Myr; at a later age, say 300 Myr, this depth approximately doubles. The mixing-in of elements produced by nucleosynthesis in the deeper layers of the star with the light elements of the stellar atmosphere may explain the peculiar pattern of abundances.

Talon et al. (2006) re-evaluated the evolutionary code (of the Montreal group) used to calculate the abundances and found that, with approximately the same parameters as used for other types of stars, the Geneva-Toulouse code leads to turbulent transport coefficients that produce

---

[14]This is a parameter describing the mass of the stellar envelope under which there is mixing between deeper layers, where products of nucleosynthesis from fusion reactions reside, and higher layers that are yet unaffected by nucelosynthesis. The location of the various layers within the star is indicated by the fractional stellar mass above a specific layer.

[15]Stellar models often use the mass as an independent variable, instead of the radius. The mass variable changes from zero at the outer boundary to the full mass of the star at the center.

abundance anomalies consistent with the observed ones for Sirius among other stars. In order to differentiate among this model and that from the Montreal group, Talon et al. concluded that much more accurate abundance determinations are required.

In the same context of chemical abundances, but based on the UV spectral measurements with the *Copernicus* satellite, Yushchenko et al. (2007) concluded that the atmosphere of Sirius A was contaminated by s-process elements. They proposed that this was produced by mass transfer from the atmosphere of Sirius B during the red giant phase.

5.3.1. Rotation

In general, early-type stars of types O, B, A, and to a certain degree F, are known as fast rotators, with equatorial velocities into the tens or hundreds of kilometers per second. The earlier the spectral type of a star is, i.e., the more massive the star is, the faster it spins. Slettebak (1955) determined an observed average rotational velocity of 210 km s^{-1} for B stars, 170 km s^{-1} for A stars, and 30 km s^{-1} for F stars. Sirius was studied extensively to determine this property. The conclusion is that the rotation of Sirius A is much slower than expected, when compared with stars of the same spectral type.

Stellar rotation is determined from the width of the spectral lines. In principle, one is unable to determine the velocity itself, but can at most derive a value for $v_e \sin i$, where v_e is the equatorial rotational velocity and i is the inclination of the stellar rotation axis with respect to the normal to the line of sight.

The high-dispersion spectrum of Sirius, among those other A stars, was first investigated for rotation by Westgate (1933). She found that the rotational speed is essentially zero, which was interpreted by her as an indication that the star is presumably viewed in a pole-on orientation. In a review of spectra and rotational velocities of B8–A2 stars, Slettebak (1954) quoted Sirius as having a zero rotational velocity (in fact, serving as a zero rotational velocity standard).

A study by Geary & Abt (1970) of A and early-F stars in the Ursa Major stellar group (in which Sirius was claimed to be a member) showed a minimum in the distribution of rotational velocities near spectral class A1; Geary & Abt explained this as the influence of Ap and Am stars in the sample; these stars were known to be slow rotators. Eggen (1950) presented information about the rotation of Sirius in his compilation of photometry for plotting the color-magnitude diagram of nearby stars. He quoted there papers from 1930 to 1933, where the rotational velocity of Sirius is given as 0 km s^{-1}. Eggen suggested, as previously did Westgate (1933), that this might be the result of viewing Sirius in a "pole-on" orientation.

Sirius is part of the compilation of rotational velocities of the A0 (B9.5 to A1) family of stars (Dworetsky 1974) where the quoted values for it are 0 km s^{-1} from Uesugi & Fukuda (1970), and 10 km s^{-1} from Bernacca & Perinotto (1970, 1971). Dworetsky concluded that the distribution of rotational velocities among the A0 stars can be modelled by a Maxwellian law only if one accepts the presence of a population of intrinsically slow rotators. This population constitutes 60% of all the slowly-rotating stars with $v_e \sin i \leq 40$ km s^{-1}. As one of the possible explanations for the slow rotator population, Dworetsky proposed that these are remnants of binary systems in which mass transfer from the progenitor of the white dwarf companion to the star that is now the primary took place, bringing about the slowing down of the A0 star.

Smith (1976) analyzed a single absorption line in the spectrum of α CMa A produced by Fe I at $\lambda 4476$ Å. This was obtained with the echelle scanner at the Coudé focus of the 2.7-m McDonald Observatory by David Lambert and it had a very high signal-to-noise. Smith used a model atmosphere from Kurucz et al. (1972) with T_{eff}=9500 K and fitted the full-width at half-maximum of the line with a rotational velocity of 17±1 km s^{-1}. Smith could not fit a definite value for the microturbulence[16] in the atmosphere of Sirius, and mentioned cryptically that *the microturbulence in Sirius is either very low or very high.* It is also possible that the line shows a measure of macroturbulence[17], perhaps at 2.5 km s^{-1}.

Milliard et al. (1977) used the high-resolution UV spectrum of Sirius obtained by the Copernicus satellite to determine anew its rotational velocity. They used only a small segment of the UV spectrum, centered on the B II $\lambda 1362.46$ Å line, in which they analyzed the profiles of the Cl I $\lambda 1363.449$ Å and the Fe II $\lambda 1362.771$ Å lines. The resulting velocities ($v \sin i$), obtained after deconvolving the line profile via a Fourier transform, were 11.4±2 and 10.6±2 km s^{-1}.

Another determination of the rotation of Sirius A was obtained by Kurucz et al. (1977). They looked at Ba II $\lambda 6496.9$ Å and measured, using an interferometric device, a projected rotational velocity $v \sin i$=16±1 km s^{-1}. Note that they assumed microturbulence values of 2 km s^{-1} and macroturbulence of 0 km s^{-1}. The radial velocity they measured for Sirius A was −8.6±0.4 km s^{-1} on January 28, 1975, as predicted for this epoch using the binary star elements of van den Bos (1960).

---

[16]This is the non-thermal velocity field in a stellar atmosphere that has a spatial scale smaller than the mean free path of a photon. In principle, this causes overall broadening of a spectral line.

[17]This is a property similar to the microturbulence (see above) that has a spatial scale larger than the mean free path of a photon. In principle, the macroturbulence represents gas bulk motions in the stellar atmosphere.

Gray & Garrison (1987) refined the classification of the early-A stars with newly-obtained spectroscopy, and compared it to Strömgren and Hβ photometry and to the effects of rotation. Their remark about Sirius, a star that is not emphasized in any way in their paper, was only that its Ca II *"K line [is] marginally weaker than that of α Lyrae"*. No specific reference was made to the slow rotation of Sirius A.

Ramella et al. (1989) surveyed all the northern ($\delta > -10°$) unevolved A0 stars listed in the Bright Stars catalog (Hoffleit & Jaschek 1982), which are known to rotate slower than 40 km s^{-1}. They quoted in their Table 2 three values of $v \sin i$ for Sirius (HD 48915): 18.6 km s^{-1} for the profile obtained by folding the red wing of the line about the barycenter of the line, 19.6 km s^{-1} for the profile obtained by folding the blue wing of the line about the barycenter of the line, and 19.0 km s^{-1} for the entire line. They used the Fe II λ 4233.167 Å line for their analysis, and the uncertaintiy they quoted was of order 3–4 km s^{-1}.

Sirius was studied by Dravins et al. (1990) with extremely high spectral resolution ($\frac{\lambda}{\Delta\lambda} \approx 130,000$) and with extremely high spectral purity[18] at the Coudé Auxiliary Telescope (CAT) of ESO using an echelle spectrometer. They presented results for four spectral lines and included the best-fitted values for $v \sin i$ and for the deduced "intrinsic" line widths (i.e., not broadened by stellar rotation and instrumental resolution). Their results are given in Table 5.1.

Dravins et al. (1990) concluded, as many others did before, that since Sirius is rotating so much slower than the average for its spectral type, this could be the result of our viewing it in a pole-on configuration, when most of the equatorial velocity would be across our line of sight. They even calculated that the inclination of the axis of rotation of Sirius to the line of sight could be at most of order 10°.

Griffin et al. (2000) used a very high dispersion spectrum of Sirius, obtained with the Mount Wilson 100-inch telescope and Coudé spectrograph at

TABLE 5.1. Rotation of Sirius A (Dravins et al. 1990)

Species	λ (Å)	v sin$i$ (km s^{-1})	FWHM (km s^{-1})
Fe I	4271.153	15.8	5.4
Fe I	4271.759	15.4	8.1
Fe I	4383.544	14.8	9.6
Fe II	4385.373	14.8	8.7

[18]The spectral purity depends on the ability of an instrument to isolate a wavelength region from nearby regions.

0.75 Å mm^{-1}, to determine a rotation of 17 km s^{-1} (16.5 km s^{-1} excluding micro-turbulence).

The conclusion from the various observations is that Sirius A is rotating much slower than most late-B to early-A stars. The different rotational velocities quoted in the literature concentrate near $\sim$16 km s^{-1}, while the general population of early-A stars shows rotation ten times faster, at $\sim$170 km s^{-1}. The observed rotational velocity is a result of Sirius being an intrinsic slow-rotator star, spinning much slower around its axis than most A stars. This might be compounded by a near-pole-on viewing angle from the Solar system. However, the maximal axial tilt calculated by Dravins et al. is very far from the inclination of the orbit of Sirius B (138°.4; Holberg 2007) while one would expect the spin and the orbit to be aligned. Thus the origin of the slow rotation of Sirius A remains unexplained.

5.3.2. Magnetic field

Sirius, like most other stars, has a magnetic field. The field is probably generated by plasma motion inside the star within the convective zone. In general, stars have dipolar magnetic fields. The magnetic lines become wrapped because of differential rotation. The magnetic field of a star can be measured by means of the Zeeman effect[19].

The magnetic field of Sirius was first measured by Babcock (1958a). Light from the telescope was split into two beams of orthogonal polarizations. The two spectra were recorded simultaneously and were compared. The sample of stars in his catalog consisted of sharp-lined A-type stars in which the detection of magnetically-split lines could be attempted. In his discussion of magnetic fields of A-type stars (Babcock 1958b) he listed Sirius as having no magnetic field at all. In fact, the catalog entry for Sirius is an upper limit at 1,000 G.

Babcock mentioned that the visibility of metallic lines in the spectra of main-sequence A stars depends on the rotational broadening of the lines; stars with small $v_e \sin i$ show a rich spectrum of weak metallic lines. The rotational widening of the lines makes the profiles shallow and broad, thus the splitting of lines becomes much harder to detect. Although Sirius is known to be a sharp-lined A star, it did not exhibit a Zeeman effect.

All the stars for which Babcock found magnetic fields were "peculiar" A-type stars (Eggen 1950). Such objects show strong spectral features of Manganese, Europium, Chromium, and Strontium. In the case of Sirius, Morgan (1932) detected the presence of Europium already in the 1930s thus the star could be expected to show a magnetic field.

---

[19]The effect consists of the splitting of a spectral line into a number of components, in the presence of a magnetic field. The degree of splitting depends on (and measures) the strength of the magnetic field. The Zeeman effect originates in a quantum physics property produced by the existence of "magnetic quantum numbers".

A possible detection of a magnetic field in Sirius was reported by Severny (1970), following the use of an observational technique used for solar magnetograms. The method scanned an exit slit across a spectral line that was projected in a Coudé spectrometer at the focus of the 2.6-m Crimean Observatory reflector and a photometer signal was recorded for each circular polarization direction. Severny (1970) reported a possible detection of a magnetic field as weak as +38±12 G from Sirius (quoted as 55±5 G in Didelon 1984).

Borra (1975) attempted to detect the magnetic field of Sirius with the Coudé polarimeter at the 200-inch reflector of Palomar observatory using the photoelectric method, expected to yield higher accuracy than the photographic one. He focused on the Fe II line $\lambda4520.2$ Å and measured its polarization with 86 mÅ steps, along with the polarization in the adjacent continuum. The upper limits he reached show that the longitudinal magnetic field of Sirius cannot be greater than a few tens of Gauss.

In this context, Didelon (1984) argued that the weak magnetic field of Sirius, as measured by Severny, fits the picture of Manganese (Mn) stars having weak magnetic fields. He explained the field as being produced by an "oblique rotator" model, in which the magnetic field is frozen in the star at an angle β from the rotation axis. It is possible that there is a small de-centering effect also, with the magnetic dipole shifted with respect to the rotation axis of the star[20].

Didelon (1984) discussed also the possibility of braking the stellar rotation, which could occur if the magnetic fields would couple between the star and the nearby interstellar medium. The braking is proportional to the surface magnetic field and to the obliquity of the field. The more oblique the field, the stronger the braking effect is ($\sim 10^6$ years are needed to brake a rotator with the magnetic field perpendicular to the rotation axis). Although he mentioned Sirius among the weak-field Am stars, Didelon did not attempt to link a possible rotational braking effect with the presence of a companion.

Takada-Hidai & Jugaku (1993) searched for a magnetic field in Sirius by analyzing pairs of Fe II lines near 4400 Å and 6150 Å. Their analysis was based on a correlation between the strength of the magnetic field and the difference in equivalent widths of the lines in the pair normalized to their average equivalent width. They obtained two values for the magnetic field, 0.0 kG using an empirical relation, and $\sim$2–2.5 kG using a theoretical relation. Only the first value fits the determinations of the magnetic field of Sirius obtained from spectro-polarimetry. The conclusion, from the different measurements described above, is that if Sirius has a magnetic field at all, it must be a relatively weak one.

---

[20] An off-center magnetic dipole, which is oblique to the rotation axis, is the accepted model for the Earth's magnetic field.

5.3.3. Gravitational Redshift and Spectra of Sirius B

Einstein's 1915 General Theory of Relativity (GR) predicted that light, just
as matter, responds to the gravitational attraction of other bodies. This
was demonstrated directly by observations of starlight deflection by the
Sun made during the total solar eclipse of May 19, 1919, and is considered
one of the experimental verifications of GR. During a total solar eclipse the
direct light from the Sun is blocked by the Moon, thus light rays from stars
seen in projection close to the edge of the Sun pass very close to a big mass
concentration while remaining visible against a darkened sky. The presence
of the Sun causes a measurable shifting of the light paths and the stars are
observed in slightly different positions (radially shifted outward) from the
locations they have when the Sun's mass is not bending their light.

The GR prediction is that any centrally condensed object would act also
upon its own light rays. These would emerge from a deeper gravitational
potential than rays from a star of the same mass but with a much smaller
density. I already mentioned that Sirius B has a high density, because its
mass is similar to that of the Sun but its radius is similar to that of the
Earth. In fact, its average density is six orders of magnitude greater than
that of normal stars. This affects the light emitted by its surface and, in
particular, the spectral lines. These will be redshifted, in comparison with
the same lines produced in the laboratory or at the surface of a normal
star, because the photons will lose energy escaping from the surface of the
dense star.

The gravitational redshift expected at the surface of a white dwarf whose
mass is M and its radius is R is (Trimble & Greenstein 1972):

$$K = 0.635\, M/M_\odot\, (R/R_\odot)^{-1}\, \text{km s}^{-1} \qquad (5.2)$$

or, as a fractional wavelength shift (Phillips 1994):

$$\Delta\lambda/\lambda \approx 74[M/M_\odot]^{4/3}[GM_\odot/R_\odot c^2] \qquad (5.3)$$

The Sun, for example, has a K value of slightly more than 0.5 km
s^{-1}. If the Sun were to be compressed to a radius of 5000-km so that it
would be equivalent in size to a white dwarf, K would become $\sim$89 km
s^{-1}. Such a gravitational redshift would move the center of the Hα line
by $\sim$2 Å from its laboratory value. As this is a direct determination of
the WD mass, requiring only a spectroscopic observation, it is one of the
more important among astronomical measurements. Alternatively, if the

WD mass is independently known, for example from a good binary solution, the gravitational redshift measurement can be used the determine the WD radius.

Attempts to detect the gravitational redshift from Sirius B started almost immediately after the magnitude of the effect was predicted by Eddington to be $\sim+20$ km s^{-1} (Eddington 1924). Sirius B is the most suitable white dwarf in which to detect this effect because of its nearness, brightness, and relatively large mass. At the same time, it is also a difficult target because of the scattered light from Sirius A that compromises spectroscopic observations. However, this is an excellent test of the method, because there are other methods to determine the mass of the WD in the Sirius system, for example, from a full solution of the binary system dynamics (Gatewood & Gatewood 1978).

The first attempts to measure the gravitational redshift from the white dwarf component of the binary system were by Adams (1924) and by Moore (1928). Adams referred to this measurement as "the fifth test of the Theory of Relativity". The two measurements yielded essentially the same value: $+19$ km s^{-1}. While Adams did not give a formal statistical error while reporting his result, the dispersion of the different measurements indicates a standard deviation of order 4 km s^{-1}. A similar error can be estimated for Moore's measurements. This gravitational redshift was confirmed by Strömgren (1926) using the shift measured by Adams and the proper radial velocity for the B component for the date of observation. Moore noted also that the spectrum of Sirius B appeared of a later type than that of the A component:

> the hydrogen lines are broader and the enhanced lines of Fe, Ti, and Sr, prominent in Sirius [A] are weaker in the spectrum of the companion... The companion's spectrum appears to be about Class A5.

The value for the gravitational redshift measured by Adams (1924) was criticized by Hetherington (1980) as indicating a lack of objectivity. He claimed that Adams was, in fact, looking for a redshift value close to that predicted (wrongly) by Eddington, based on the then-known properties of Sirius B. The same measurement, by Adams (1925) was also explained away as a non-discovery by McCrea (1972) as being produced by *the light of Sirius A after reflection by Sirius B*.

Greenstein et al. (1985) and Wesemael (1985) have both criticized Hetherington's (1980) assertion. Greenstein et al. explained in great detail the difficulty of obtaining a reasonable spectrum of the WD in the Sirius system. They re-examined the original Mt. Wilson 100-inch telescope photographic plates and discovered that the WD spectrum was indeed heavily contaminated by light from the main-sequence star. Not only the Balmer lines, normally present in the spectrum of a DA white dwarf, were contaminated by the deep absorption lines of the A star, but the spectrum of Sirius B seemed to show also metallic absorption lines unlike those seen in any other white dwarf; these were produced by the same scattered light from Sirius A.

Wesemael (1985) reconstructed the situation encountered by Adams when he tried to measure the gravitational redshift of Sirius A. He combined in various proportions the spectrum of a white dwarf with the scattered light from the spectrum of a main-sequence A star using theoretical spectra for both. His simulation shows that for a ~30% contribution of scattered light, one would measure a line shift similar to that seen by Adams (1925). Therefore, the value measured then was similar, just by chance, to that predicted by Eddington.

Wiese & Kelleher (1971) studied laboratory plasmas at different electron densities in order to determine how are the Balmer line profiles affected by this variable. They studied, in particular, the regime $N_e=10^{16}$–10^{17} cm^{-3} and the temperature range from 10,080 to 12,600 K. Their relevant finding was that all Balmer lines were highly asymmetric, with shifts that were consistently toward longer wavelengths (apparent redshifts) when measured by the method of Greenstein & Trimble (1967). This effect, of redshifts caused by the high electron density, must be compensated for in all measurements of gravitational redshifts.

Greenstein et al. (1971) analyzed the profiles of Hα and Hγ in Sirius B obtained with the 200-inch Hale reflector, after apodizing[21] the entrance aperture of the telescope in order to move the diffraction pattern of Sirius A off the image of Sirius B. They obtained a gravitational redshift of +89±16 km s^{-1} and, by fitting model atmospheres of hydrogen-rich white dwarfs, derived log g=8.8 (assuming T_{eff}=28,000 K this comes out as log g=8.0). Correcting for the Stark[22] shift, reduced this gravitational redshift value by ~8 km s^{-1}.

A comparison between spectroscopically-determined masses, obtained by fitting line profiles to derive surface temperatures and surface gravities by comparison with model atmospheres, and masses determined from gravitational redshifts for 35 DA white dwarfs was made by Bergeron et al. (1995). Although Sirius B was not one of the stars included in their sample, this study is relevant because it showed the gravitationally-determined masses to follow to within 1σ the spectroscopic determinations of the mass. Therefore, the use of gravitational redshifts to measure white dwarf masses is a relevant and accurate method.

[21] A process of purposely changing the input intensity profile of an optical system by masking part of its entrance aperture.

[22] Quantum effect discovered in 1913 by the German physicist Johannes Stark (1874–1957) in which spectral lines separate into a number of components with slightly different wavelengths under the influence of an electrical field. This is the electric analogue of the Zeeman effect of line splitting under the influence of a magnetic field.

The most accurate (to date) determination of the effective temperature and of the gravitational redshift of Sirius B is by Barstow et al. (2005). The observations were in the visible spectrum and were targeted to analyze the Balmer lines. Barstow et al. used the HST to observe Sirius B on February 6, 2004, after carefully rolling the spacecraft to move the diffraction pattern spikes from the A component away from the white dwarf. This allowed the acquisition of spectra of Sirius B as free as possible from the scattered light of the A component. Figure 5.7 shows the field with the diffraction spikes of Sirius A moved away from the B component.

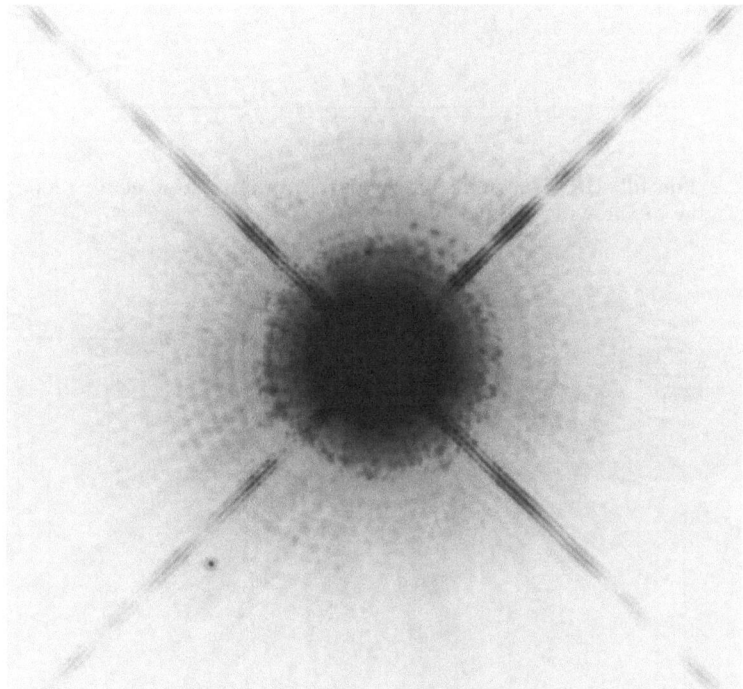

Figure 5.7. This *HST* image of the Sirius system, with exquisite angular resolution, was obtained by rolling the spacecraft to distance the Sirius A spikes from the white dwarf. Sirius A is the bright stellar image at the center, shown here as negative, and Sirius B is the faint star near the bottom left spike (NASA image).

The 52×0.2 arcsec STIS spectrometer slit was aligned horizontally in this image, sampling the WD, the surrounding scattered light background, and the diffraction spikes from A. The full spectrum obtained with the G430L is shown in Figure 5.8 and details of the blue part of the spectrum, with a superposed model fit, are shown in Figure 5.9.

Barstow et al. (2005) used the HST and STIS with the G750M grating to provide sufficient resolution to explore the core of the Hα line. Since the

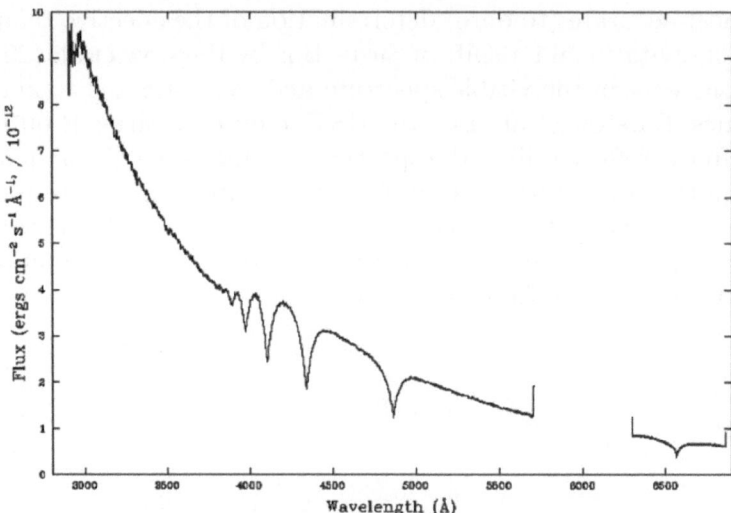

Figure 5.8.   The full HST spectrum of Sirius B from Barstow et al. (2005) shows an exquisite display of the wide Balmer lines.

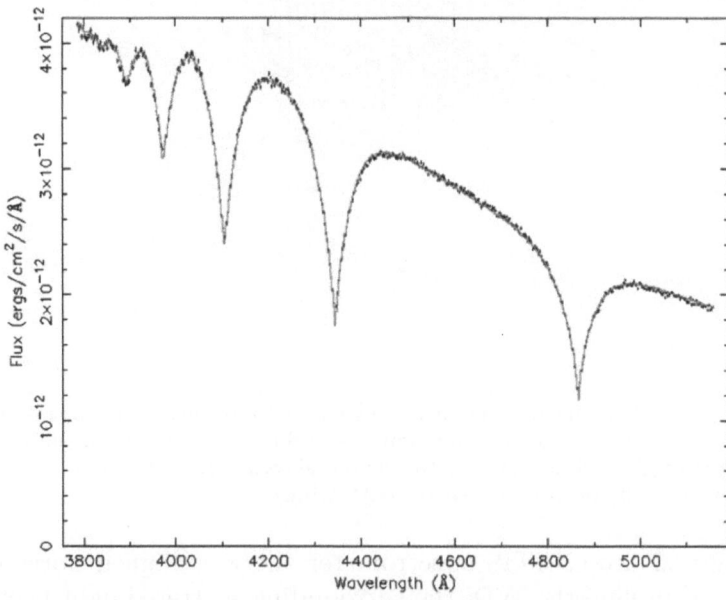

Figure 5.9.   The blue part of the HST spectrum of Sirius B from Barstow et al. (2005) shows the excellent fit of the observations (points) to the theoretical spectrum (continuous line).

entire line was imaged onto ∼300 pixels on the STIS detector, Barstow et al decided to use only the line core (±7 pixels) to evaluate the gravitational redshift. After correcting the measured value of the shift for the space velocity of the Sirius system barycenter, and for the orbital velocity of Sirius B around the barycenter, the authors found a redshift of 80.42±4.83 km s^{-1}, fitting well the mass-radius relation for white dwarfs (see below Figure 6.5.

5.4. Conclusions

Sirius is a bright star both because it is intrinsically bright, in comparison with stars of the solar neighborhood, and also because it steadily moved closer to the Sun in the last millions of years. While most of the visible light comes from the A component and it is reasonably easy to measure, the light from B is always affected by scattered light from A. This scattered light, present in any optical system, affects also the spectra of the B component. Sirius A has different metal abundances than most A-type stars. It is a slow rotator and, as will become clear later, there is no good reason for that. It also has a very weak, perhaps negligible, magnetic field. One of the big accomplishments of space-age astronomy is the measurement of the gravitational redshift of Sirius B.

Chapter 6
Modern non-optical observations

Modern astronomy is no longer relegated to the optical domain. The expansion of the data collection into additional spectral domains brought about a revolution in how we see the stars, the galaxies, and the space between them. The use of space-borne telescopes extended not only the angular resolution in imaging to almost undreamed-of values, but also allowed the astronomers access to the X-ray, the ultraviolet (UV), the infrared (IR), etc. All these enriched very much the material available to confront observationally measured parameters of astronomical sources with theoretical models and understand their nature, their development, and evolution.

6.1. Infrared

A major step in understanding the immediate environment of stars was the discovery by the *Infrared Astronomy Satellite (IRAS)*[1] of dust around early-type stars. Tasked with the mapping of the infrared (IR) sky, this instrument operated long enough to scan the entire sky more than three times. Its observation of Vega (α Lyrae), which is an A0 star similar to Sirius, showed a significant excess of infrared emission with respect to that expected from such a hot, almost black-body object (Aumann et al. 1984). The analysis showed that there are dust grains around Vega and that these should be much larger than the typical interstellar grains, rather similar to the particles of the zodiacal dust cloud in the Solar System. The distribution of the dust producing the 60 μm emission (in a disk, some 20 arcsec wide around the star), and the fitted temperature for this dust (approximately 85 K), showed that significant amounts of particles could be found around an A0 star.

Aumann (1985) conducted a systematic search for similar sources of IR emission near early-type stars and found twenty other Vega-like stars. He concluded that millimeter-sized dust could be found near early-type stars and that the dust distribution there must be either optically-thin, because

---

[1] The IRAS satellite was built and operated by the US, UK, and the Netherlands. It was launched in 1983 to map the IR sky and operated for 10 months.

the dust bolometric brightness was always $\approx 10^{-4}$ of the star's bolometric luminosity, or the spatial configuration of the dust must be very flat.

Chini et al. (1990) extended the search for dust to lower temperatures by observing in the sub-millimeter regime with the IRAM 30-m telescope. They looked at Vega and at four other nearby stars, one of which was Sirius, at 800, 870, and 1,300 μm. Their measurements show that Sirius has only a very modest IR excess up to 870 μm, but no excess whatsoever at 1,300 μm. The excess could be explained either as a statistical fluke, due to the low flux values, or as the presence of some dust heated to 50 K. Because of the low reliability of the sub-mm excess, if it exists, Chini et al. were not able to derive the mass of the dust. They remarked that any dust in the Sirius system could not be located in a flat distribution, because that configuration would have been perturbed by the presence of the white dwarf companion (unless the dust would have been co-planar with the WD and far from it).

Song et al. (2001) studied a volume-limited sample of 200 A-type stars and found that 13% of these could be Vega-like by exhibiting IR excesses. They estimated the ages of these stars from their Strömgren colors and found that they may be either very young ($\sim$50 Myr) or rather old (1 Gyr). It seemed that the presence of excess IR emission from an A-type star is not related to the age of the star, apart from the fact that younger stars show more emission than older ones.

Habing et al. (2001) used the *Infrared Space Observatory (ISO)* satellite, launched by ESA in 1995 and operated till 1998, to measure a sample of 84 main sequence stars of types A to K at 60 and 170 μm. They found that about half the stars younger than 400 Myr showed FIR excesses that could be attributed to the presence of dusty disks. From this, they concluded that most stars reach the main sequence after being formed with a dusty disk, which dissipates after $\sim$400 Myr. They rejected Sirius from their sample, because they eliminated all binaries with a distance between companions smaller than 1 arcmin. Habing et al. noted that the presence of a companion star does not preclude the possible presence of a disk; both Jupiter and Saturn in our Solar System have dust disks yet they have many satellites as well.

Despite the rather small number of measurements, it seems to be fairly certain that the α CMa system does not exhibit a significant FIR excess, as shown by other A stars. This should be interpreted as a lack of significant amounts of circumstellar material and is relevant in the context of the evolutionary stage of the binary system.

Kervella et al. (2003) used ESO's Very Large Telescope Interferometer (VLTI) to measure the apparent diameter of Sirius A in the K band (2.2 μm) and found $(5.936 \pm 0.016) \times 10^{-3}$ arcsec for a uniformly illuminated stellar disk, and $(6.039\pm0.019)\times10^{-3}$ arcsec for a limb-darkened disk. They mentioned that the measurement cannot rule out a slight asymmetry of the disk of Sirius, or the presence of a disk around the star that could produce ∼1% of the stellar flux. Similarly, these observations could also not rule out the presence of an M5 dwarf orbiting close to Sirius A, as proposed by Benest & Duvent (1995).

6.2. UV and EUV measurements

The aim of UV and EUV observations of the Sirius system was, among others, to accurately determine the effective temperature (T_{eff}) of Sirius B. This is a difficult task in the optical because of the nearness of the A component, whose light overloads, in general, the detection of the WD at wavelengths longer than ∼1,500 Å. An early attempt to measure T_{eff} by Greenstein et al. (1971) used detailed fits of the Hα and Hγ Balmer lines to yield T_{eff}=32,000 K. Somewhat later, Koester (1979) found T_{eff}=22,000 K, while even later papers narrowed the range down to 26,000±2,000 K (Holberg et al. 1984, Paerels et al. 1988, and Kidder et al. 1989).

However, these attempts were done with data obtained in the optical domain. The white dwarf is easier to detect at shorter wavelengths, because of its higher effective temperature that enhances the contrast between it and the A component. UV spectral observations can also reveal absorption lines produced by intervening interstellar matter; from this one can determine the distribution, density, and physical properties of ISM clouds in the line of sight to the observed target.

The necessity to establish the accurate T_{eff} of Sirius B arose because of the need to accurately determine its radius (through the luminosity and specific emissivity $\sigma = \kappa T_{eff}^4$ of a black body). The measurement of the radius has implications on the possible equation of state of white dwarfs, the relation defining the density, temperature and pressure of a degenerate electron gas, as will be discussed below. Sirius B is an important stepping stone in this relation, because it is one of the more massive white dwarfs and one that is very close to us, thus easier to study.

One of the first UV observations of Sirius was by the astronauts Lovell and Aldrin in November 1966, from the *Gemini 12* spacecraft (Spear et al. 1974). The UV spectra were obtained on film, with a hand-held 70-mm camera equipped with a 22-mm aperture, 73-mm focal length UV lens covering a field of view of 30°. The astronauts used an objective-grating that provided spectra covering the range 2,200–4,400 Å at a nominal dispersion of 183 Å mm^{-1}. The two Sirius spectra showed the spectrum expected from a hot Am star and the spectral lines blocked ∼15% of the stellar continuum in the 2,650–2,910 Å range. A plot of the spectrum derived from the *Gemini 12* observations is shown in Figure 6.1.

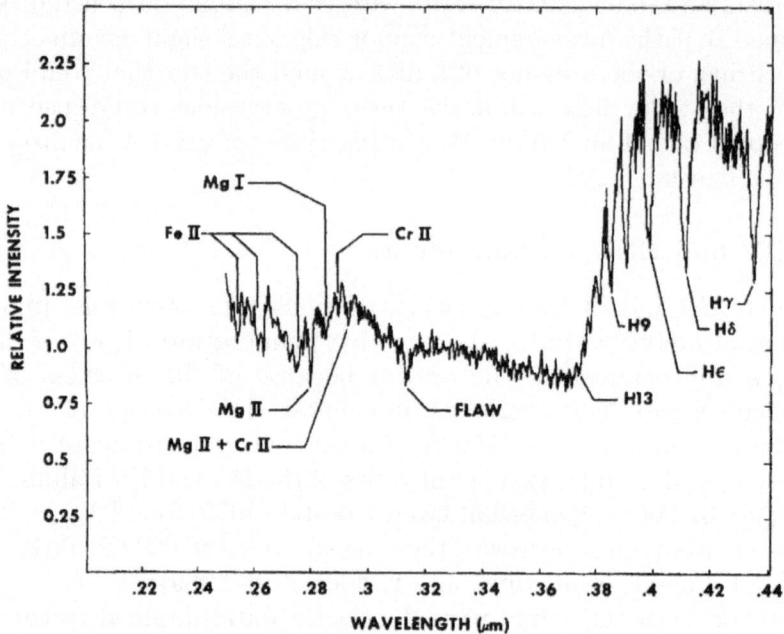

Figure 6.1. The first UV spectrum of Sirius obtained by the astronauts Lovell and Aldrin from the *Gemini 12* spacecraft (from Spear et al. 1974).

At about the same time that the manned spacecraft obtained UV spectra of Sirius, on November 21, 1966, an Aerobee rocket lofted a UV spectrometer to perform UV observations of the brighter stars (Stecher 1970). Four stars were observed with the 13-inch telescope mounted in the rocket nose cone, one of which was α CMa. The spectrum obtained by this experiment covered the range 1,150–3,200 Å in two segments, depending on the detector used in the spectrometer. Stecher mentioned that the most prominent feature in the recorded spectrum was the very broad Lyman α line.

About 6 months after this flight, another Aerobee rocket lifted a far-UV electronographic type spectrometer equipped with objective gratings to an altitude of 193-km, in order to observe the spectra of some early-type stars (Carruthers 1968). This instrument covered the range 950–1,400 Å. Sirius was one of the targets and was the only star from which the zero order image was recorded, presumably because of its brightness. The results did not help much in understanding the EUV emission of Sirius, partly because the payload was damaged during re-entry and could not be recalibrated.

A number of attempts were made during the Apollo-Soyuz Test Program (ASTP) to detect the Sirius system at 100 Å (Shipman et al. 1977) and at 300 and 570 Å (Cash et al. 1978). These attempts were not successful and resulted only in upper limits. Upper limits were obtained also from an Aerobee rocket flight on March 23, 1973 (Riegler & Garmire 1975), when an attempt was made to measure the flux from Sirius and from a few other bright stars at three EUV bands defined by filters: 85–240 Å, 140–430 Å, and 140–700 Å.

The *Copernicus* satellite observed the Sirius system on January 9, 1975 (Savedoff et al. 1976). The observations consisted of three different data sets: a scan of the 1,100–1,448 Å range with ∼0.16 Å steps, spot coverage of the 1,000–1,344 Å range with ∼0.02 Å steps, and continuous coverage from 2,200 to 2,896 Å with ∼0.32 Å steps. A similar observation, performed 11 days later, included only the A component in the spectrometer slit; this allowed its contribution to be subtracted from the combined A+B spectrum obtained earlier, thus yielding the net spectrum of Sirius B. This "first-time" observation showed that the spectral energy distribution was rising at shorter wavelengths, beyond the Lyman α line. The result allowed Savedoff et al. to calculate an effective temperature of 27,000±6,000 K for Sirius B.

The very high spectral resolution provided by the V1 spectrometer onboard *Copernicus* allowed Kondo et al. (1978) to study the Mg II resonance doublet (2,795.5 and 2,802.7 Å) in the spectrum of Sirius, along with a few other nearby stars. This yielded, for the first time, the determination of the column density of Mg II in the direction of Sirius at a distance of 2.67 pc. The derived value was $(2.44\pm0.73)\times10^{12}$ cm^{-2}.

The UV spectrum of Sirius obtained by the *Copernicus* satellite was also used by Boyarchuk & Snow (1978) to determine the abundances of some heavy elements. Among other findings, they detected for the first time the presence of Cd, and set upper limits to the abundances of Hg, Pt, and Tc. The limits set for Hg and Pt were close to the solar abundance values, implying that Sirius has no enhanced abundance of these elements, although the abundances of other metals seemed to be enhanced.

Very close after the launch of the *International Ultraviolet Explorer (IUE)*[2] satellite, it was used to obtain a short-wave UV spectrum of Sirius B (Böhm-Vitense et al. 1979). This yielded a spectral energy distribution in the continuum that was in agreement with that of theoretical white dwarf models with $T_{eff} \approx 26,000\pm1,000$ K, log g=8.65, and R=5.08×10^8 cm.

Far-UV observations of the Sirius system were performed by Holberg et al. (1984) with the *Voyager 2* spectrometer (UVS: Broadfoot et al. 1977). For stellar sources, the spectral resolution was 18 Å and the sensitivity was optimized for the 800–1,200 Å region. In fact, both *Voyagers* observed

---

[2]The highly successful *International Ultraviolet Explorer* spacecraft, launched in 1978 and operated until 1996, carried two spectrometers covering the UV range from 1,100 to 3,300 Å.

Sirius on numerous occasions, because Sirius was chosen to be a calibration standard for the ultraviolet spectrometers. The spectrum analyzed by Holberg et al. collected 11,700 s of exposure on November 3, 1981. Note, though, that the UVS instruments did not have concentrating optics and the entrance aperture to the grating is only a rectangle defining the field-of-view ($0°.1 \times 0°.87$). However, the sky region where Sirius is located does not contain UV-bright stars that could confuse the observations. Holberg et al. succeeded in compensating for the flux contributed by Sirius A, which is mostly longward of Lyman α (1,216 Å), by using a scaled spectrum of α Lyr (Vega), which is also a nearby A0V star and does not have a WD companion, and information collected in a similar way for the white dwarf CD −38°10980. They concluded that most of the light shortward of ∼1,000 Å was produced by Sirius B.

The *Voyager* UVS observations of Holberg et al. (1984) restricted the definition of a rather large region in the temperature-radius relation for the white dwarf component, which was defined also by *IUE* observations and data coming from the X-ray domain. The most likely T_{eff} was found to be 27,300 K (the formal effective temperature was 27,000±1,000 K). In particular, the UVS observations put a firm upper limit on the temperature at 28,000 K.

A spectrum of the Sirius system in the extreme ultraviolet is part of the *Extreme Ultraviolet Explorer (EUVE)* Stellar Spectral Atlas. The detection was possible because of the extreme brightness of the object. Otherwise, its relatively low effective temperature would have prevented its detection in the EUV domain. The detected EUV spectrum is the combination of the Wien part of the Planck black-body SED and the reduced opacity of HI toward shorter wavelengths. The flux drops from the Lyman edge at 912 Å to a minimum near 350 Å, then rises to a peak near 150 Å. The photon flux (photons cm^{-2} s^{-1} Å^{-1}) at 150 Å was as high as that at 800 Å.

An even better *EUVE* spectrum was obtained by Holberg et al. (1998) and is shown here as Figure 6.2. Its analysis showed that the spectrum might contain Helium lines, indicating a Helium abundance ratio of He/H=1.8×10^{-5}. This is somewhat surprising, because theoretical calculations of diffusion time scales for Helium show that the expected abundance ratio should be $\leq 10^{-8}$. This interpretation, of the presence of Helium in relatively large amounts in the atmosphere of Sirius B, is in need of confirmation with spectra of higher signal-to-noise.

Holberg et al. (1998) combined the *EUVE* observation of Sirius with reprocessed archival *IUE* observations, from which they derived the spectral energy distribution (SED) from 80 to 2,100 Å. The SED was fitted with model atmospheres to yield a temperature T_{eff}=24,790±100 K and a surface gravity log $g = 8.57 \pm 0.06$. These values, when combined with the

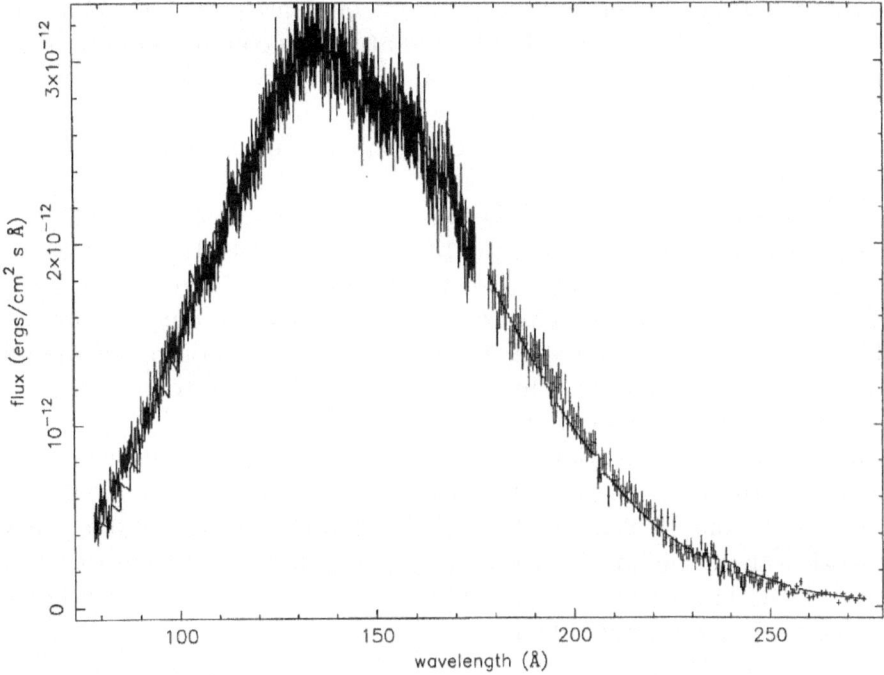

Figure 6.2. The extreme ultraviolet spectrum of Sirius shows clearly the contribution of the white dwarf. This is a combination of the short-wave and long-wave spectra obtained by the *EUVE* satellite during a 92,430 s observation performed at the end of November 1996 (from Holberg et al. 1998).

Hipparcos parallax and with the astrometric mass derived by Gatewood & Gatewood (1978), yielded a mass of 1.034±0.026 M$_\odot$ and a radius of (8.4±0.25) × 10^{-3} R$_\odot$, or 5,600-km.

Holberg et al. (2004) reported results from *FUSE* observations in the Lyman line region shortward of Lyα. The observations demonstrated that the spectrum was not contaminated by light from Sirius A and that, apart from the Hydrogen lines themselves, no other features were detectable.

6.3. High energy observations

The X-ray emission from Sirius was observed for the first time by the *Astronomical Netherlands Satellite (ANS)* in April 1975 (Mewe et al. 1975). *ANS* carried an X-ray telescope consisting of a parabolic mirror with a proportional counter at its focus. The telescope was sensitive to X-rays in the range of ~0.2–0.284 keV. However, the counter was also sensitive to ultraviolet light; in order to account for this contribution, and to subtract it properly from the measurement in the direction of Sirius, the authors repeated the measurement with a filter blocking the UV photons.

The measurement of the X-ray emission from the Sirius system was attempted during an Aerobee rocket flight on March 24, 1973 (Patterson et al. 1975), where the system was looked at with a collimated proportional counter. The field of view was $5° \times 10°$ and the observations were performed above an altitude of 160-km. This observation could not account for the UV photons from the system, thus was only able to put an upper limit to the X-ray flux from Sirius at 10^{29} ergs s^{-1}.

The detection of soft X-ray photons in the band from 0.20 to 0.28 keV coming from Sirius implied an X-ray luminosity of $(9.1\pm1.6)\times10^{27}$ ergs s^{-1} at a distance of 2.66 pc. The observations of Sirius were repeated with *ANS* and no variation was detected on a time scale of days. Mewe et al. (1975) argued that the source of the X-rays is in a hot corona surrounding the white dwarf component. This was supported by D'Antona & Mazzitelli (1978), resulting from model atmospheres they calculated.

Further positive detections of Sirius in the X-ray domain were by the *HEAO A-2* satellite (Lampton et al. 1979) and by the *Einstein Observatory* satellite (Giacconi 1979). The *Einstein* observation, in particular, showed that both stars in the Sirius system are X-ray sources. The High Resolution Imager (HRI) instrument showed that ~99% of the X-ray flux originates from the white dwarf component.

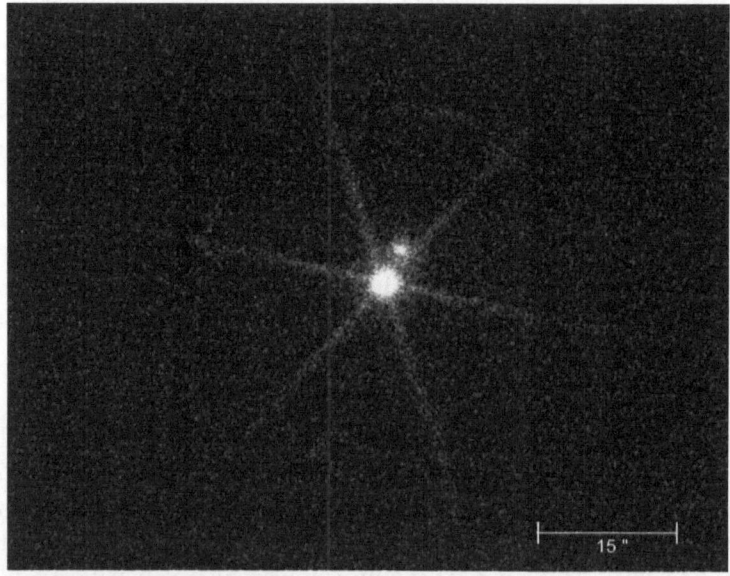

Figure 6.3. This *Chandra* X-ray image of the Sirius system, with exquisite angular resolution, shows two X-rays sources; the brighter one is Sirius B, the white dwarf, and the fainter image is Sirius A(NASA image).

The soft X-ray spectrum of Sirius, obtained by the *HEAO 1 A-1* experiment, was studied by Martin et al. (1982) in conjunction with other X-ray, EUV and UV measurements. They fitted those to model atmospheres and concluded that the best fit is with a photospheric effective temperature in the range 27,000 K$<$T$_{eff}$ $<$28,500 K and an upper limit to the Helium abundance $\frac{n_{He}}{n_H} \leq 3 \times 10^{-5}$.

The *Chandra* satellite obtained an X-ray image of Sirius, shown here as Figure 6.3.

Beuermann et al. (2006) established Sirius B as one of the three primary soft X-ray standards by cross-calibrating *Chandra* LETG+HRC observations with those from the *ROSAT* PSPC and the short-wavelength *EUVE* spectrometer. They used model spectra of hot white dwarfs and fitted Sirius B with T$_{eff}$=24,923±115 K for a fixed log (g)=8.6.

6.4. Basic stellar parameters

A first (unsuccessful) attempt to measure the apparent angular diameter of Sirius A was by Chacornac (1864). The angular diameter of Sirius was measured for the first time with the Narrabri Observatory intensity interferometer[3] (Hanbury Brown et al. 1974) at (5.89±0.16) ×10^{-3} arcsec.

Code et al. (1976) derived basic astrophysical information about Sirius A by determining its effective temperature in an empirical way. They used the Narraabri apparent angular diameter of the star and the absolute flux distribution from a combination of experimental measurements in different spectral bands. The total flux was derived from ground-based optical spectrophotometry and from UV spectro-photometry using the *OAO-2* satellite and its spectrometer. The resultant spectrum ranged from 1,100 to 8,080 Å , with an IR photometric extension up to the N-band (8–12 μm). Figure 6.4 reproduces the spectral energy distribution (SED) they derived for Sirius from the UV and optical spectro-photometry.

The resultant flux, at (114.3±4.4) ×10^{-6} erg cm^{-2} s^{-1}, translates into an emergent flux at the stellar surface of (5.61±0.37) ×10^{11} erg cm^{-2} s^{-1}, which yields an effective temperature of T$_{eff}$=9,970±160 K for Sirius A.

Ten years later, Davis & Tango (1996) used another Australian facility, the Chatterton Astronomy Department's stellar interferometer, to measure the apparent diameter of Sirius. This instrument was a double sidereostat on a North–South baseline separated by 11.4 m, directing beams of light through evacuated pipes into a central laboratory. An adaptive optics system operated at this specific location to reduce the wavefront error induced by atmospheric turbulence to ∼0".1 and to introduce the required beam delays.

---

[3]Intensity interferometers use the measured intensity observed by two or more detectors to correlate the arrival times of photons.

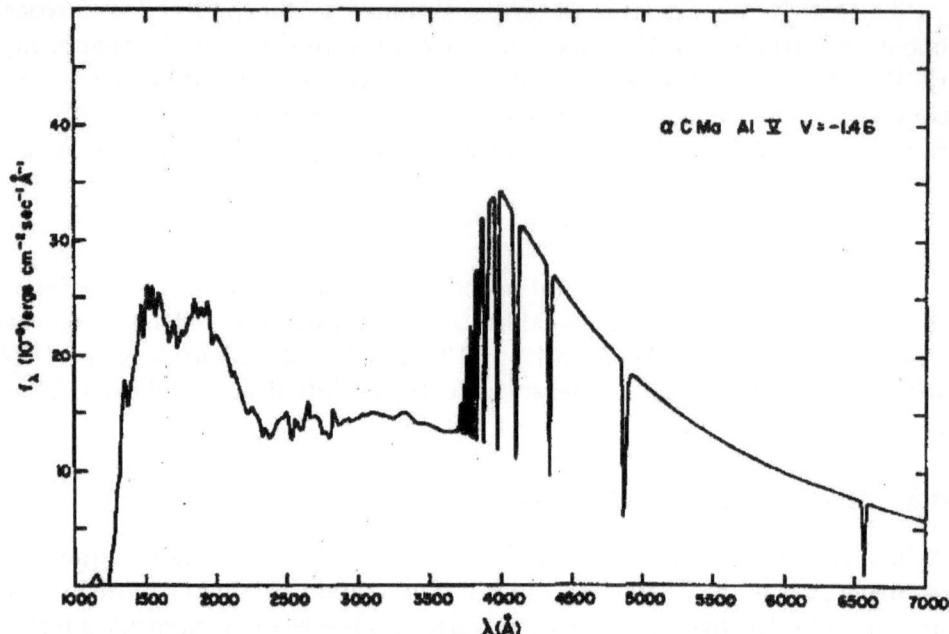

Figure 6.4. The spectral energy distribution of the Sirius system, from 1,100 to 8,080 Å (Code et al. 1976), shows clearly the contribution of Sirius B at wavelengths below ~2,200 Å. The spectrum of the A-type primary is characterized by strong and deep Balmer absorption lines. Using more modern observations it may be possible to produce a better-quality SED.

Using this instrument, Davis & Tango (1996) measured Sirius on six different instances in 1986 and found a mean diameter (assuming uniform illumination of the stellar disk) of $(5.63\pm0.08)\times10^{-3}$ arcsec, identical within the errors to the previous measurement of Hanbury Brown et al. (1974). The effective temperature of Sirius A seemed, therefore, to be firmly established at slightly less than 10,000 K. The radius of Sirius A, determined to within ~1%, is therefore $(1.121\pm0.016)\times10^{6}$ km.

Kervella et al. (2003) used their near-infrared results from the VLTI interferometric measurement of the diameter of Sirius quoted above, together with a set of evolutionary models, to determine the history of Sirius A. In particular, they determined that the age of Sirius A is $(200–250)\pm12$ Myrs.

The excellent quality HST spectra of Sirius B obtained by Barstow et al. (2005), while minimizing the contamination by scattered light from Sirius A, allowed the fitting of a model atmosphere that provided, together with unpublished values for the white dwarf of $M=0.978\pm0.005$ $M_{\odot}$, a well-determined radius $R=(8.64\pm0.12)\times10^{-2}R_{\odot}$, and a gravitational redshift of 80.42 ± 4.83 km s^{-1}. The good fit of these parameters to the theoretical mass-radius relationship for WDs is shown in Figure 6.5.

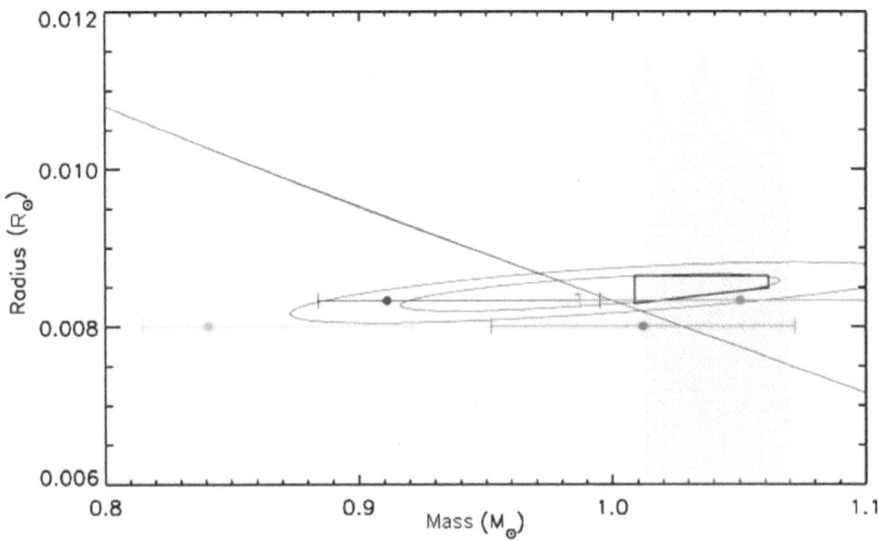

Figure 6.5. The derived mass and radius of Sirius B from Barstow et al. (2005) matches quite well the theoretical relation for carbon-core WDs. The ellipses are 1σ and 2σ confidence contours from Holberg et al. (1998). The error bars represent fits to observations obtained with different gratings.

6.5. Conclusions

We have seen how modern observations succeeded in improving the knowledge about the orbit of Sirius, from which the derivation of masses for the two components was obtained. The redetermination of the gravitational shift of Sirius B, and improvements in characterizing its brightness, spectral energy distribution, etc. offered additional checks.

The number of observational questions regarding this system was significantly reduced. What still remains on the table are questions about the origin of the chemical peculiarities of Sirius A, the reason for its slow rotation and the lack of a significant magnetic field, the existence of a possible dust disk or faint third companion in the system, the higher-than-expected Helium abundance of Sirius B, and the origin of the X-ray emission. These, as well as the established parameters, have to be explained by a comprehensive formation and evolution model of the Sirius system.

Of special attention are some oblique claims of a possible modification of Sirius A due to its evolution in a binary system (due to possible common-envelope evolution during the red giant phase of Sirius B): the slow spin of Sirius A when compared with most A-type stars, its metallic-line character, etc. These will be discussed further below.

Chapter 7
The neighborhood of Sirius

Stars do not live in isolation. They are members of systems of stars with the largest local unit being the Milky Way galaxy. The space between the stars contains interstellar matter (ISM) composed of neutral and ionized atoms, molecules, and dust grains. This material can be cold with temperatures of a few to a few tens of degrees, tepid at a few thousand degrees, or hot reaching millions of degrees. The density of the ISM also varies, from much less than one particle per cubic cm to many 10^4 cm^{-3}.

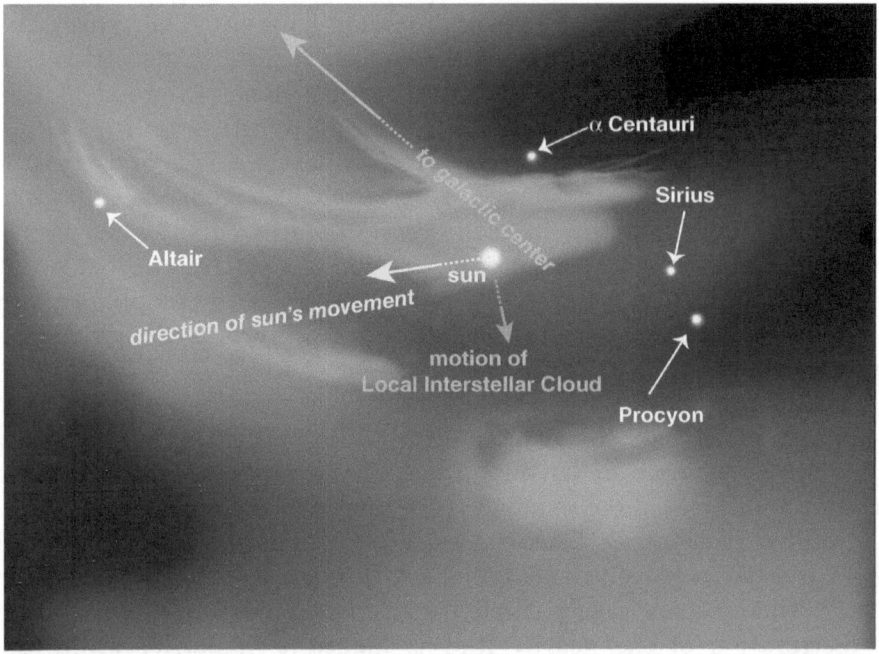

Figure 7.1. Artist impression of the distribution of very local ISM (Source: Priscilla Frisch and American Scientist).

The ISM between us, the observers, and a distant star may modify the observed properties of the target star. In particular, this could be the case for Sirius and such very local ISM might have modified the way Sirius appeared to observers in antiquity. In order to account properly for these modifications, one needs to study the ISM properties in the line of sight to our target.

7.1. Interstellar matter

The Sun and Sirius are located in the same "interarm" region of the Milky Way (MW) galaxy. The spiral structure of the MW is determined from the distribution of the hot and blue young stars, from the mapping of HII regions since these are sites where the young stars form, and from the location of the young stellar clusters. Additional structural information is obtained from the mapping of the free-electron density of the Galaxy using pulsar dispersion measures and from the distribution of neutral Hydrogen and of molecular clouds of the ISM. In the direction of the Galactic Center, which is some 8 kpc away, we observe the Scorpio-Sagittarius (Sco-Sag) arm of the MW at a distance of only $\sim$2 kpc. In the opposite direction, toward the anti-center and also some 2 kpc away, we see the Perseus (Per) arm. The Sun and its immediate neighborhood, including Sirius, are sited near the inner edge of the Local Arm (Orion-Cygnus). The third galactic quadrant, from l=220° to l=250°, is deficient in ISM. This region is also known to harbor a deformation of the galactic plane (Alfaro et al. 1991), as found from mapping the distribution of open clusters.

The location between galactic arms, which are loci of intense star formation, implies that the Sun and its immediate stellar neighbors have been removed for some time from the very energetic regions where stars are currently being formed. The interarm region is relatively devoid of interstellar matter (ISM; see Figure 7.2); this was shown by mapping the reddening of stars with known distances (Lucke 1978, Frisch & York 1983, and Paresce 1984).

The clearing of ISM from the general Solar neighborhood is also due to the activity of the Scorpio-Centaurus super-bubble. This inflated some 15 million years ago an asymmetric cavity around the location of the Sun, presumably as a consequence of strong winds from young stars and supernova explosions. A back-tracing of the solar motion indicates that the Solar System travelled for millennia through the third quarter of the Galaxy, where the average particle density is less than 0.02 cm^{-3}. In fact, the distribution of the nearby ISM is highly asymmetric: while to the Galactic center direction the neutral Hydrogen density is n(HI)$\approx$0.1 cm^{-3}, to the anti-center it is $\sim$0.003 cm^{-3}. Moreover, the neutral matter in the anti-center direction, for the first 200 pc, is concentrated only in the first $\sim$5 pc.

As Sirius is a nearby star, it is important and perhaps also relevant to the evolution of this binary system, and to the way it was seen from

180

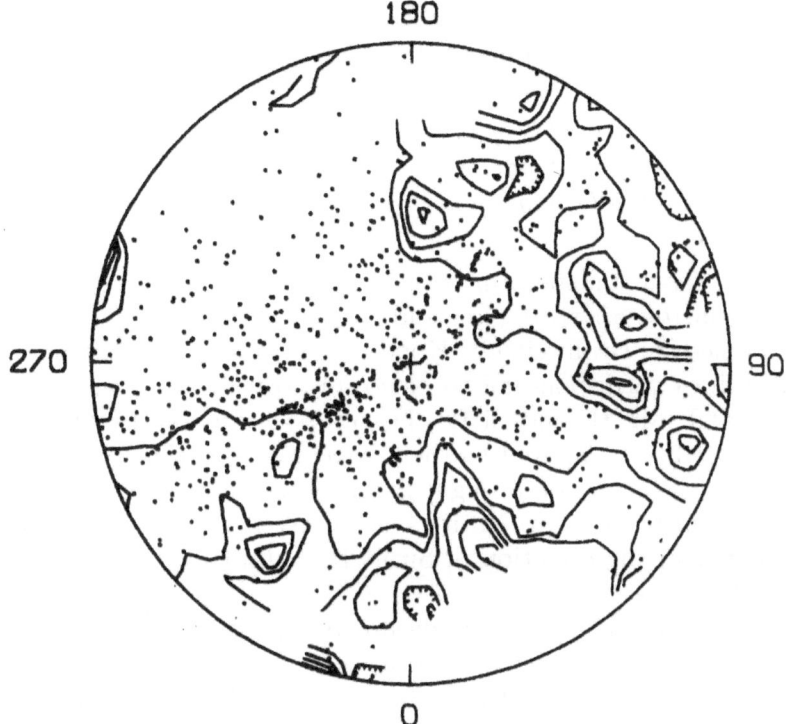

270   90

0

Figure 7.2.   The distribution of the local interstellar material up to a distance of 500 pc from the Sun. This is Figure 4 from Lucke (1978) and is based on the E(B–V) color excess of ~4,000 O and B stars plotted here as dots. The lowest contour in this plot corresponds to $N(HI)=5\times10^{20}$ cm^{-2}.

the Earth in ancient times, to consider what lies between us and it. This is because evidence accumulated in recent years regarding the extremely fragmented nature of the local interstellar medium (LISM), where the line of sight to Sirius shows the presence of a relatively dense (>10 cm^{-3}) cloud of neutral Hydrogen (Zank & Frisch 1999). The immediate environment of the Sun is (almost) the same as that of Sirius, thus understanding it could help in the understanding of some of the strange findings about Sirius.

The interstellar gas is generally associated with interstellar dust grains, with a "typical" Hydrogen-to-dust ratio of $N(HI)/E(B–V)=5.8\times10^{21}$ cm^{-2} mag^{-1} (Bohlin et al. 1978). Here $N(HI)$ is the column density of neutral Hydrogen and E(B–V) is the selective extinction between the B and V bands produced by the dust particles. Therefore, the presence of neutral Hydrogen in the line of sight to a star such as Sirius could imply also the presence of dust, and that could cause reddening of the starlight. In fact, the first physical explanation for the possible redness of Sirius in antiquity, proposed by Herschel in a letter published in 1839, was just that, the influence of interstellar extinction. Herschel wrote (quoting from Ceragioli 1995):

> *It seems much more likely that a red colour should be the effect of a medium inferred than that in the short space of 2000 years so vast a body should have actually undergone such a material change in its physical constitution.*

The high-dispersion spectrum of Sirius B, obtained by the *IUE* satellite through the narrow aperture (to avoid contamination by the light of Sirius A), was studied by Bruhweiler & Kondo (1982, 1983). They detected only interstellar lines corresponding to electronic transitions from the ground state (C I, O I, Mg II, and Si II). Despite the apparently wider profiles of these lines when compared to their appearance in the spectra of other white dwarfs, this detection brought Bruhweiler & Kondo to the conclusion that the absorption lines are entirely of interstellar origin. This indicated that the LISM in the direction of Sirius B contains not only dust grains, but also diffuse gas.

The realization that ISM clouds are present in the immediate Solar neighborhood, within 20 pc, was already proposed more than 40 years ago by Munch & Unsold (1962). They realized that between α Oph, at a distance of 18 pc, and the Sun there is an ISM cloud, and that this cloud has structure on scales smaller than 1 parsec. This inference was strengthened in later years by high-resolution spectroscopic observations toward the nearest stars, which showed many velocity systems (e.g., Ferlet et al. 1986 and Lallement et al. 1994). Velocity systems are identified when the high-resolution spectra of a background star show, instead of a single absorption line from an element present in the ISM, a number of absorption lines from the same element separated by relatively small velocity differentials. The velocity difference between the systems is usually much larger than the sound speed in rarefied neutral gas, which is a few kilometers per second.

Lallement et al. (1994) used the Goddard High-Resolution Spectrograph[1] (GHRS) to study the line of sight to Sirius with very high resolution ($\frac{\lambda}{\Delta\lambda} \simeq 90,000$). The analysis of the GHRS observations, obtained under the same HST program, revealed a stellar wind emitted by Sirius and the existence of a small cloud of very hot gas, possibly an envelope around the system, at a temperature of 10^5 K (Bertin et al. 1995).

An analysis of the very high resolution spectra towards Sirius has been presented by Izmodenov et al. (1999). HST observations with the GHRS

---

[1]The Goddard High-Resolution Spectrometer is an instrument that operated on the Hubble Space Telescope and provided high spectral resolution and high signal-to-noise spectra in the ultraviolet of single object point sources.

and performed in the UV domain were reported also by Hébrard et al. (1999). Such information, published in many papers, serves to reconstruct the distribution of the ISM in the nearby Milky Way.

The distribution of different clouds of matter in the solar neighborhood was described by Frisch (1995, see also the artist impression in Figure 7.1). She showed that the Solar System is embedded in a warm, tenuous, partly-ionized cloud with the following parameters: $T \sim 10^4$ K, $n(HI) \sim 0.1$ cm^{-3} and $n(HII) \sim 0.22$–0.44 cm^{-3}. Her view of the LISM is that of a turbulent, stormy system of clouds. In a very picturesque way, Frisch described an extended and expanding system of clouds which she dubs *the squall line*, in analogy with the cloudy front of a storm. The squall line consists of the cloud complexes within $\pm 60°$ of the direction to the Galactic center. The local system of ISM clouds she called *the local fluff*. It represents material of physical thickness 3–6 pc to the anti-center direction, and was defined by Frisch (1995) as

> *the gas observed by the Copernicus satellite in front of the nearest stars, at $d < 4$ pc.*

Frisch (1997) analyzed the distribution of the "nearby" interstellar matter, within 500 pc. She collected information about ISM clouds and found that the molecular clouds (mainly CO) are essentially at rest with respect to the Local Standard of Rest. The Sun moves at 19.5 km s^{-1} with respect to the kinematical LSR (defined as the average velocity of stars with spectral types A through G).

Frisch concluded from the information she collected that the environment in which the Sun finds itself embedded changes with time, and that this may influence the immediate distribution of ISM near the Sun, even that within the Solar System. In particular, the "local fluff" complex of ISM clouds may have been completely different 200,000 years ago than what is observed today. Frisch identified spikes in the ^{10}Be abundance in the Antarctic ice, detected by Sonett et al. (1987) and dated at 33,000 and 60,000 years ago, as created by encounters of the Solar System with dense interstellar cloud material. She argued that these encounters caused an influx of ^{10}Be-rich ISM into the Solar System, with some of the material reaching the Earth.

The claims that interstellar material enters continuously our Solar System are based on Lyman α and He I $\lambda 584$ Å photons backscattered off interplanetary material, on various types of interstellar ion measurements, and is supported by the spacecraft detection of ISM dust in the outer regions of the system (eg., Krüger et al. 2007) and by the detection of interstellar meteors.

The item that is more relevant to the present discussion is the distribution of the Local Interstellar Medium (LISM) from the Sun in the direction of Sirius, and the mode by which this could have changed during the last millennia. According to the motion-mapping of the nearby stars, Sirius lies "downstream" from the Sun, thus the Sun is the first to encounter new LISM material. This material is subsequently translated into the line of sight to Sirius as seen from the Sun. Further motion takes Sirius out of the LISM cloud.

An attempt to trace the structure of the LISM was done by Génova et al. (1990). They used absorption lines of Mg II detected with the Long Wavelength spectrograph of *IUE* to measure the Mg II column density seen in absorption in the spectra of some late-type stars within 30 pc of the Sun. The values measured in the direction of α CMa were found to be $(2.04\text{--}3.80) \times 10^{-7}$ cm^{-3}. The results for the entire sample of stars indicate, according to Génova et al., that the Local Cloud (LC) cannot extend further away than ~ 2 pc from the Sun. One of the possibilities they put forward is that the kinematics observed in the LC may be produced by an evaporating small cloud, which is denser and cooler than the general LISM.

Frisch (1995) supposed that the local fluff, seen also in the direction of Sirius, could be an extension of the squall line system. In Table IV of her paper she gave the column density of the material in the direction of Sirius: $N(HI) \approx (1.1 - 5.0) \times 10^{18}$ cm^{-2}, as derived from a compilation of sources [mainly N(H) values derived of from soft X-ray observations in the line-of-sight to Sirius B; Paerels & Heise 1989]. Her Table V gives even more detail, using the HST GHRS data of Lallement et al. (1994). Two systems were identified there, one at ~ 20 km s^{-1} and another at ~ 14 km s^{-1}. There was a different column density of Mg II to these clouds; $N(Mg$ $II) \approx (1.7 \pm 0.2) \times 10^{12}$ cm^{-2} to the first cloud and $\sim (8.5 \pm 1.5) \times 10^{11}$ cm^{-2} to the second.

The analysis of the GHRS observations, assuming a typical Magnesium-to-Hydrogen ratio (by number) N(Mg II)/N(HI+HII) of 3.24×10^{-6} for the local fluff, yielded a Hydrogen column density of 8×10^{17} cm^{-2} and an average density of 0.09 cm^{-3}; the column density measurements from the EUVE observations of Sirius B yielded values about one order of magnitude greater [$N(HI) \approx (4 - 13) \times 10^{18}$ cm^{-2}]. Frisch (1995) mentioned that the discrepancy between these values was not understood. By the way, a similar discrepancy in the column density of Hydrogen is apparent in the spectrum of α Cen A at 1.3 pc. There, an anomalous D I line was observed at -8 ± 2 km s^{-1} (Landsman et al. 1984).

Frisch (1995) presented a model that allows the dating of the squall line and of the local fluff. According to it, the local fluff evaporated off the surface of ISM clouds in the Sco-Oph association a few (1–5) Myrs ago. The squall line, on the other hand, is a much more recent feature; it was formed about 400,000 years ago from a cloud of homogeneous gas.

The second cloud observed toward the Sirius system, which has a velocity of $\sim$14 km s^{-1}, is a puzzle according to Frisch (1995). This is because this feature, blue-shifted in respect to the local fluff velocity, is also very cold (T=1,000±3,000 K) and has a larger turbulent velocity than the other cloud: ξ=2.9(+0.1, −0.5) km s^{-1}. The fluff cloud in the direction of Sirius, for comparison, has T=7,600±3,000 K and ξ=1.4(+0.6, −1.4) km s^{-1}. Similar values were measured for the line of sight to Capella (Linsky et al. 1993) and to Procyon (Linsky et al. 1994, 1995). The local fluff component shows both Mg I and Mg II absorption components; the aberrant cloud does not show Mg I at all.

The observations of Linsky et al. were obtained with the GHRS. Two-epoch observations of Capella (α Aurigae, a binary with G 1 III and G8 III components, at a distance of 12.5 pc), allowed the disentangling of the intrinsic stellar profiles and of the interstellar profiles, from which the temperature of the LISM was derived to be 7,000±500 K, with a micro-turbulence parameter of 1.6±0.4 km s^{-1}. The analysis of Procyon (α CMi, an F 5 IV-V star at 3.5 pc) revealed two velocity components separated by 2.6 km s^{-1} (20.8±1.5 and 23.4±1.5 km s^{-1}); the line of sight (LOS) to Procyon had 2.4 times more mean Hydrogen column density than that to Capella.

The results of the analysis of two lines of sight toward stars close to the Sun but at large angular separations demonstrated clearly the existence of very small structures in the LISM. The second absorption component seen in the spectrum of Procyon is different from that seen in Sirius (-6.2 km s^{-1} relative to the Local Cloud), even though the two stars are in the same general part of the sky. The differences between the Capella LOS and the Procyon LOS, separated by 52° in the sky, are explicable by placing Procyon near the edge of the LC and having Capella fully out of the LC. The Sirius cloud, which must be nearer than 2.7 pc, is not visible in the spectrum of Procyon thus it must be smaller than 1.2 pc. The same situation exists for the Procyon cloud. Therefore, the two clouds must be at most $\sim$1 pc in size. The neutral Hydrogen has a volume density of $\sim$0.1 cm^{-3} and its measured "non-thermal" motions may be, instead, a reflection of shear in the LISM.

Combining the different motions, Frisch (1995) reached a conclusion that the solar motion is parallel to the local fluff, with the Sun being embedded in the fluff. Sirius is located close to the anti-apex of the solar motion and is on a tangential line of sight to the local fluff. This LOS samples the interface between the local fluff and the postulated high-temperature coronal substrate of the ISM, as well as sampling the solar wake in the ISM. Frisch concluded that the Solar System first encountered the local fluff during historical epochs, 2,000–8,000 years ago.

The basic solar motion weighted more heavily by the radial velocities of stars of the most common spectral types (A, gK, dM) in the solar vicinity is $\sim$16.5 km s^{-1} toward the galactic direction l=53° and b=+25°. This corresponds to a transversal distance of 16.5 pc per million years. The relative motion with respect to the local fluff takes place at $\sim$26 km s^{-1} (5.4 a.u. per year or $\sim$0.05 pc/2,000 year). This is supersonic motion, and the Sun is expected to leave a wake made up of solar wind plasma and ISM Hydrogen containing heavier elements. It seems that in the upwind direction the LISM is compressed and forms a "Hydrogen wall". This was detected by analyzing the diffuse Lyman α emission in the low-dispersion UV spectra obtained by the *Voyager* 1 and 2 spacecraft (Quémerais et al. 1995).

The presence of dust grains in the LISM, specifically in the very nearby region from which dust grains penetrate into the Solar System, was discussed by Frisch et al. (1999). The experimental data were collected during the last decades by the *Ulysses, Helios, Galileo* and *Cassini* spacecraft, the first while it was out of the ecliptic plane and the others close to the ecliptic and mostly between 0.3 and 3 a.u. In the anti-center direction, the ISM has a gas-to-dust mass ratio of $\sim$500, while the measurements within the Solar System, for the interstellar matter flowing through and measured by the *Ulysses* and *Galileo* spacecraft, yield a ratio of only 94^{+46}_{-38}. These values, however, represent only the gas-to-dust ratio for the relatively large ISM grains, with radii greater than 0.2 μm (grain masses of 10^{-13}–10^{-18} kg). A reconsideration of these values was done by Frisch & Slavin (2003). The issue of interstellar dust entering the Solar System was reviewed by Krüger et al. (2007).

A three-dimensional reconstruction of the Local Interstellar Cloud (LIC) was attempted by Redfield & Linsky (2000). Their model is based on a spherical harmonics decomposition of the spectral information from 16 lines of sight. The spectra were obtained with the GHRS on HST, and the presence of the LIC was inferred from Hydrogen column densities derived from Deuterium column densities. This allowed the derivation of the HI column density, because the ratio N_{DI}/N_{HI} is approximately constant and optical depth effects, which are very severe at the center of the HI Lyman α line, are reduced considerably when observing the Deuterium Lyα. The spectral information was augmented with observations of WDs by EUVE (comparison of theoretical SEDs with observed SEDs) and by spectral information about the Ca II lines at the LIC velocity in 13 additional lines of sight.

The result of the modelling was that the LIC could be described as an elongated feature, whose axis of symmetry points to $l \approx 315°$ if viewed from the north Galactic pole. This calculation showed, as did that of Frisch (1995), that the Sun is located just inside the LIC. However, the Redfield & Linsky (2000) model had the LIC as a single entity with a minimum thickness of 4.7 pc, while Frisch considered it to be made up of many thin filaments, cloudlets, and fluffy structures.

Ferlet (1999) reviewed the ideas about the LISM and presented a picture of the distribution of various clouds (Figure 7.3). This shows that the Sun is located near the edge of a small cloudlet of approximately constant density and about 1 pc in size, and that there is a definite interface between it and a similar but different small cloud around Sirius. These cloudlets are apparently embedded in the very low density, hot, Local Bubble.

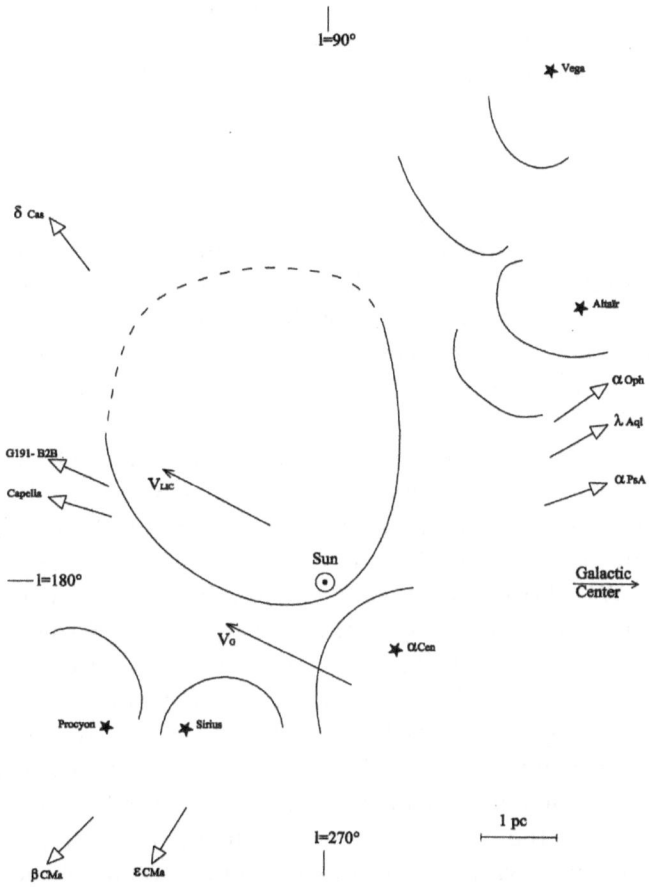

Figure 7.3. The distribution of the local interstellar material in the immediate neighborhood of the Sun. This is Figure 2 from Ferlet (1999) and shows the presence of different environments for the Sun and for Sirius. The vectors point to more distant stars.

Hébrard et al. (1999) studied the LOS to the two components of the α CMa system with the HST GHRS. At the time of the observation the two stars were separated by 4 arcsec (∼10 a.u. projected distance) and this is the divergence of the two lines of sight. Their observations, being superior

TABLE 7.1. ISM clouds in the line of sight to Sirius

Component	Blue cloud	LIC
Temperature [K]	3,000 (+2,000/−1,000)	8,000 (+500/−1,000)
Turbulent velocity [km s^{-1}]	2.7±0.3	0.5±0.3
Radial velocity	11.7±1.5	17.6±1.5

to the previous ones obtained by Lallement et al. (1994), recovered the two-cloud structure of the line of sight. Fitting 19 interstellar metal lines in the spectra allowed the derivation of the cloud parameters shown in Table 7.1.

The characterization of the two LISM clouds given in Table 7.1 is essentially that of Lallement et al. (1994). Whereas the metal lines present a picture of the LISM similar to that derived for other lines of sight, the result Hébrard et al. (1999) obtained for the two Lyman α lines is somewhat different. This result was difficult to derive because the interstellar lines were blended with the photospheric lines of the two stars. The interesting result was the detection of an absorption excess in the blue wing of the Lyman α line of Sirius A, and an absorption excess in the red wing of the Lyman α line of Sirius B.

Because of the small scales involved, Hébrard et al. (1999) explained the Lyman α differences as being caused in the Sirius system itself. Specifically, they interpreted this as a stellar wind emitted by Sirius A (already proposed by Bertin et al. 1995) and as the core of the Lyman α photospheric absorption from the WD, both superposed onto the interstellar feature. The latter assertion fits the expected gravitational redshift of Sirius B. Excluding the stellar-produced lines, Hébrard et al. found column densities of HI of $(2.4\pm1.0) \times 10^{17}$ cm^{-2} and 4.0 (+1.5/−1.0)$\times 10^{17}$ cm^{-2} produced by the "Blue Cloud" and by the LISM between us and Sirius, respectively.

Note also that Hébrard et al. (1999) could not confirm the presence of an additional absorption component in the Lyman α profile, which Izmodenov et al. (1999) interpreted as being caused by the neutralized and compressed solar wind in the downwind direction. An attempt to model the heliospheric absorption by Hydrogen and Deuterium in the direction of Sirius and of other five nearby stars was done by Wood et al. (2000).

Linsky et al. (2000) developed a new methodology to analyze the three-dimensional structure of the LISM. They used, primarily, the column density of Deuterium in different lines of sight along with N(HI) measures toward a number of hot WDs and B-type stars. Three hot WDs with both kinds of measurements were used to cross-calibrate the methods. They also assumed in the analysis that the LISM along each line of sight, which moves

with a certain velocity, has a uniform density of 0.1 cm^{-3}. From the pre-
liminary analysis, Linsky et al. concluded that it is possible that the LISM
and, in particular, the Local Interstellar Cloud, does not vary strongly on
small scales. The only exception among the lines of sight they tested was the
region near $l=220°$, where they proposed the formation of sharp structures
in the LISM due to the photoionizing radiation from ϵ CMa[2].

Linsky et al. (2000) located the Sun in the LISM cloud in the direction
of the Galactic center hemisphere, which is equivalent to the "squall line"
proposed by Frisch (1995). The Sun, according to them, is very close to
the boundary between the Local ISM cloud (LIC) and the G cloud, which
lies less than 0.054 pc away in the direction of α Cen. This distance will
be covered by the Sun in less than 3,000 years, and it is possible that
the two clouds, the LIC and the G cloud, are now in contact through a
magneto-hydro-dynamic shock front (Grzedzielski & Lallement 1996).

Reports of interstellar matter reaching the Earth and observed as radar
or optical meteors were presented by Hawkes & Woodworth (1997) and by
Baggaley & Neslušan (2002). The issue was reviewed by Baggaley (2004).
These particles are much rarer than the regular meteors and only the largest
would produce optical meteors. The spacecraft detections, and the reports
of interstellar particles reaching the Earth, strengthen the case for the presence of nearby interstellar dust that could affect the appearance of nearby
stars, including Sirius.

With reference to the extremely local structure of the LISM, I mention
also a puzzling finding mentioned by Krüger et al. (2007): the historical
records of spacecraft impacted by dust particles identified as "interstellar"
remained approximately homogeneous up to 2004, indicating that the local
dust is uniform on scales of at least 50 a.u. However, during the third orbit
of *Ulysses* around the Sun, in 2005, the direction of the dust impacts shifted
significantly southward, by at least 30° with respect to the arrival direction
of the interstellar helium flow. The reason for this direction change was
not understood at the time the Krüger et al. paper was written; it could
be related to the detailed structure of the very nearby LISM and it could
reflect on the appearance of Sirius as seem from the Earth.

7.2. Very small LISM structures in the Milky Way

One of the more interesting features of the distribution of the ISM is its
small-scale structure, which could affect the Sun-Sirius line of sight. This
has already been hinted at above, in the context of nearby lines of sight.
Here I discuss this issue from the point of view of fast changes in the
character of ISM absorption lines.

[2] Adhara; a 1.50 mag blue-white B2Iab supergiant star in a binary system at a distance
of 132 pc.

Frail et al. (1994) showed, from multi-epoch 21-cm line absorption against high-velocity pulsars, that significant opacity changes occur within a 1.7-yr interval. This implies that the ISM shows very small-scale details, even down to typical dimensions of 5–100 a.u. The largest variation in this short period, seen in the direction of PSR 2016+28, was $(4.5\pm0.1)\times10^{20}$ cm^{-2}. The ubiquitous character of the variations shows that a significant fraction of the ISM, perhaps 10–15%, belongs to the very small-scale structure. These structures were seen first by Dieter et al. (1976) in VLBI observations of extragalactic sources and were attributed to ISM cloudlets of $\sim$1 a.u. in size and densities of order 10^5 cm^{-3} comparable to molecular cloud regime, i.e., column densities of $\sim 1.5 \times 10^{18}$ cm^{-2}.

Remarkably, small-scale structure was revealed also by optical absorption line work at extremely high spectral resolution. Meyer & Blades (1996) observed with a resolution of $\sim$0.4 km s^{-1} and discovered significant spatial variations of the interstellar Na I D and Ca II K lines in the spectrum of μ Crucis, a binary consisting of two B stars separated by 38".8 (6,600 a.u. projected distance). The most striking finding was the presence of a deep and narrow Na I absorption component in one of the stars, which was completely absent in the other star. This line would have been wide if it would have originated in the stellar photosphere, and showed a significant column density of 7×10^{10} cm^{-2}. This indicated the presence of a Solar System sized cloudlet in the direction of one of these stars but not in the direction of the other. Watson & Meyer (1996) showed that similar discrepancies in ISM line strengths are visible in 17 other double-star systems, with projected separations from 480 to 29,000 a.u.

Price et al. (2000) discovered a time-varying Na I absorption line in the spectrum of δ Orionis A (Mintaka). This is an eclipsing binary composed of an O9.5II star and a B0.5III star in a 5.73 day orbit showing a partial eclipse. Note that this discovery required the use of an ultra-high resolution spectrometer, with R=880,000. The absorption line profile indicated a rather low temperature ($T_K \leq 348$ K) and a small turbulent velocity ($v_t \leq 0.35$ km s^{-1}). The changing depth of the absorption line, from a previous observation in 1994 to the reported one in 1999, while the velocity of the line remained constant, indicating the presence of a small $\sim$100 a.u. ISM cloudlet in the line of sight to the system.

Lauroesch et al. (2000) reported fast variations in the Na I absorption toward one component of the system HD 32039/40 (B9V+B8V). The time period of 4.25 year revealed significant temporal variation in the Na I column density, indicating the presence of a small-scale ISM structure of order the system proper motion (15–21 a.u.).

An additional finding, another line of sight in which time-variable absorption lines produced by ISM were identified, was reported by Crawford &

Price (2000). This is in the direction of κ Velorum (Markab), a spectroscopic binary star with a period of 116.65 days that is 165 pc away in the direction of the Vela supernova remnant, itself at $\sim$350 pc distance. The strength of the K I line in the direction of this star increased by 40% in only 6 years. The line is only 0.6 km s^{-1} wide and may even be a blend of two narrower components, only one of which probably changed in these 6 years. This indicates, according to Crawford & Price, that κ Vel is entering a dense sheet or filament of ISM that is probably only some a.u. wide.

Gry & Jenkins (2001) analyzed high-dispersion spectra of ϵ CMa obtained with the HST GHRS and with the IMAPS[3] instrument on-board the ORFEUS-SPAS mission that flew in late-1996 (Hurwitz et al. 1998). The high resolution spectra allowed the detection of three main ISM components: one at 17 km s^{-1}, a second at 10 km s^{-1}, and a third at -10 km s^{-1}. The first two components were those already detected in the spectrum of Sirius and have the highest column densities. Gry & Jenkins derived neutral hydrogen column densities from the OI and NI lines to find $\sim 4 \times 10^{17}$ cm^{-2} for the first component and $\sim 3 \times 10^{17}$ cm^{-2} for the second.

Heiles (1997) worked out a scenario to explain the observational findings indicating very small-scale features in the ISM. Apart from explaining the evidence in favor of the existence of these features, he also mentioned that in no case (either of HI or of optical absorptions) was any molecular species associated with the small-scale features. Heiles also mentioned a possible problem posed by such features, if they are ubiquitous and have properties similar to those adopted as for the average ISM: the interstellar extinction. If these structures maintain the same ratio of dust-to-gas as generally accepted for the ISM [$N(HI) \approx 2 \times 10^{21} A_V$; Bohlin et al. 1978] each would produce $\sim$0.01 mag of extinction. The overlap of many such features would cause the Galaxy to become essentially opaque, whereas we know this not to be the case.

One of the properties of dust in the ISM is that, in the presence of large-scale magnetic fields, it not only attenuates the starlight, but it also causes its polarization because of the alignment of the dust grains by the galactic magnetic field. This property was used by Tinbergen (1982) to study the distribution of the polarization in the direction of stars within 35 pc of the Sun. He concluded, from a study of $\sim$180 stars that, overall, the local dust content is very low, with $A_V \leq 0.002$. It is not likely that within the 35 pc diameter volume there is a uniform magnetic field, since changes in the direction of polarization are observed. This supports the idea of small structures in the LISM.

---

[3]IMAPS was an objective-grating echelle spectrograph designed to record spectra from $\sim$950–$\sim$1,150 Å of bright, early-type stars.

Tinbergen (1982) used α CMa to determine the instrumental polarization. As this is only one out of the 15 stars used in the derivation of the instrumental polarization, a decision whether the light from Sirius itself is or is not polarized could not have been seriously affected by its use as a calibrator. Tinbergen found that stars within the nearest 10 pc show, on average, higher polarization than stars further away. There is a possibility that this additional polarization could be caused by a very local dusty cloud. Tinbergen also identified a *Local Patch* of dust extending from l=350° to l=20° and from b=−40° to b=−5°, which was at least 30° in diameter. The magnetic field in this patch appears to be very uniform, as witnessed by the ∼uniform polarization pattern. The general direction to the *Local Patch* corresponds to that of the higher ISM extinction.

The *Local Patch* of "special" dust is certainly located no further than 20 pc away and could even be adjacent to the Solar System. The dust content of the Local Patch is not extremely high; only producing A_V ≈0.01 mag. Tinbergen (1982) speculated that the *Local Patch* could correspond to a small "cold core" of a 1 pc diameter cloud, no further than 5 pc away, because the star 36 Oph shares the polarization properties of the *Local Patch*.

Frisch & York (1986) posed the possibility that the Tinbergen patch could be clumpy; in this case a clump may even have a density as high as ∼1,000 cm^{-3} and could cause a few magnitudes of extinction for the objects it would obscur. If such a cloud encompassed at some time in the past the Solar System, its effects on the Earth would have been very visible, since its particle number density would be much higher than that of the (visible) Zodiacal cloud. One might wonder, therefore, what its effects would have been if such a cloud would have been located between us and Sirius. Note, though, that Sirius is located almost diametrically opposite to Tinbergen's *Local Patch* of dust. The influence of the *Local Patch* on Sirius may take place sometime in the future, given the direction of the Solar and Sirian motions in the Galaxy.

A catalog of polarization measurements of 1,000 stars within 50 pc of the Sun was prepared by Leroy (1993a), and an analysis of its contents was published by Leroy (1993b). Since most of the stars in this catalog are much more distant than Sirius, it is not possible to draw a specific conclusion as to the presence or absence of a small cloud in the direction of α CMa. One object referred to by Leroy (1999), HD 131334, shows significant polarization even though it is relatively nearby (51 pc using the *Hipparcos* parallax) and at high galactic latitude (+61°); this again might indicate the presence of a small and rather local dust cloud.

In conclusion, it seems that the immediate neighborhood of Sirius, of which our Solar System is a member, may harbor small ISM clouds that contain dust. Some of these clouds might even be close to the line of sight to Sirius now or may have been so in the recent past. Such clouds could have affected the appearance of Sirius as seen from the Solar System in antiquity and, because of their small size, are not doing this today. A complex dynamical situation exists between the small LISM clouds and the stars; in principle, this could have affected the way stars would have appeared to an Earth-bound observer.

7.3. Stars in the neighborhood of Sirius

Stars are not alone in space but are formed in pairs, triplets, and groups. A significant fraction of the evolution of the more massive stars happens while they are sharing the same motion as the other stars with which they were formed. The stars "evaporate" from the mass of co-eval stars because of their "thermal" velocity with respect to the mass of stars that formed together, and because of the influence of the complex gravitational field in which they live. This is primarily the gravitational potential of the Galactic disk, modulated by that of any local mass concentrations that happen to be nearby.

Historically, Sirius was considered to be the leading star in its group, which is called "the Ursa Major group" (UMa group) but which was labelled by Eggen (1958) as the "Sirius group". The core of this group are the stars in the Big Dipper and all the members form a moving group (MG). A study of the space velocity vectors of A-type stars (Eggen 1960), based on accurate parallaxes, proper motions, and radial velocities, showed that the Sirius group stands out in a plot of U (the velocity vector component away from the galactic center) vs. V (the velocity vector component in the direction of the galactic rotation). Even before Eggen's identification, the group was described by Roman (1949).

The Sirius MG was studied by Wegner (1981), who measured radial velocities of the bright primaries in 12 common proper-motion binaries containing a white dwarf. He concluded that the group was the youngest one among his sample, and that some of the white dwarfs in it could have evolved from stars as massive as 5 $M_\odot$.

Cayrel de Strobel (1981) studied the stars within 10 pc from the Sun and hotter than 4,000 K (including Sirius). These stars are also the nearest neighbors of Sirius. The study compared stars from a previous determination of metallicity $[Fe/H]$ (Cayrel de Strobel et al. 1980), which were contained in the catalog of nearby stars (Gliese 1969). The study showed that there was no significant difference in metallicity among these stars in comparison with the Sun, supporting the idea of co-evolution for this sample of nearby stars.

In the pre-*Hipparcos* age, one of the more complete kinematic studies of this group was by Soderblom & Mayor (1993). Their study used very accurate radial velocities measured with the CORAVEL machine[4], along with astrometric and spectroscopic information for solar-type stars in the UMa group. Soderblom & Mayor (1993) identified 37 members of the group, including Sirius, and determined that they can be considered as a cluster with an age of 0.3 Gyr and a mean metallicity [Fe/H]=−0.08±0.09.

Eggen (1998) reconsidered the Sirius "supercluster", a group of stars streaming together in nearby space, using results from the *Hipparcos* mission. The nucleus of this grouping of stars lies in Ursa Major, the stars slightly precede the Sun in the direction of the Galactic rotation, and they all have a low space velocity ($\sim$18.5 km s^{-1}). The plot of the stars in the Sirius supercluster in Figure 7.4 shows the absolute magnitude (determined using *Hipparcos* parallax values), against the B–V color of each object.

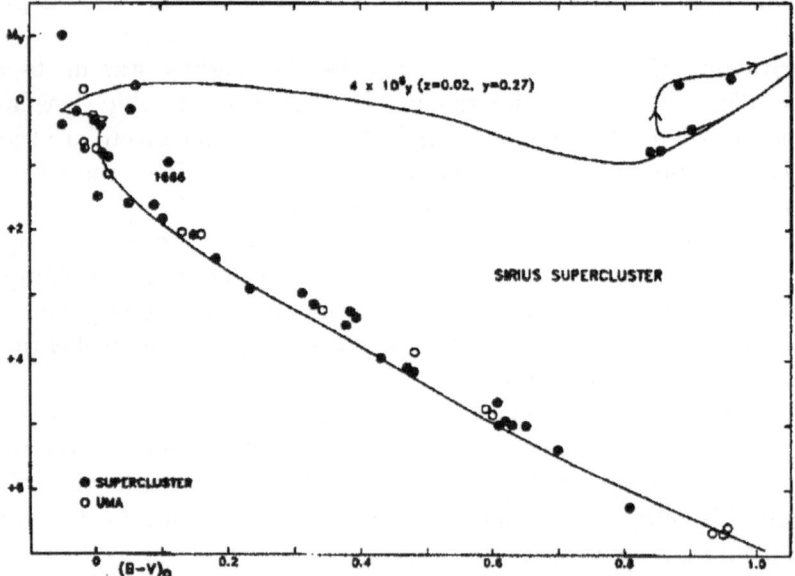

Figure 7.4. The color-absolute magnitude diagram for stars in the Sirius supercluster, after Eggen (1998). All stars but Sirius line-up properly with the "classical" isochrone for an age of 4×10^8 yr. Sirius itself is the uppermost point in the left side of the diagram and is off the general relation.

---

[4]The CORAVEL instrument correlated the spectrum of a star with a spectral template of a star of a similar spectral type by hardware. The correlation was performed by mechanically moving the optical template (a mask of the spectral lines) and the maximal correlation was detected photo-electrically.

Eggen (1998) compared the location of the stars in the diagram with the theoretical predictions for a single stellar population 400 Myr old and with abundances Y=0.27 and Z=0.02, from Castellani et al. (1992). His plot shows that all the stars line up reasonably well on the diagram, supporting the idea of a single-age population. The one outstanding object is Sirius, which seems too bright and too blue at the top-left corner of the figure to belong to this group. This is despite the kinematic arguments that argue in favor of this association.

Asiain et al. (1999) studied the Sirius MG as one among a number of similar groups. They concluded that this MG is quite extended and is probably made up of a number of subgroups, while giving its age as ∼500 Myr. Castro et al. (1999) studied the Ba abundance in stars belonging to the UMa moving group. They concluded that a probable explanation of the Ba over-abundance is the formation of those stars in a molecular cloud with a peculiar chemical abundance. Castro et al. suggested that the molecular cloud had been enriched prior to the formation of the stars they analyzed by a few AGB stars, one of which might have been Sirius B.

Olano (2001) studied the motion of the Sirius supercluster back in time without putting any special emphasis on Sirius itself. He concluded that a supercloud of 2×10^7 M$_\odot$ and a radius of 400 pc was the precursor of the Sirius supercluster, of Gould's Belt, and of the Local arm. The Sirius supercluster was born 500 Myr ago in the supercloud, rotating in the Galactic sense. The supercloud entered a major spiral arm 100 Myr ago. The interaction with the Galactic shock and the subsequent gas streaming generated braking and gas compression, which led to the drifting out of the stars of the Sirius supercluster from the gas and to the formation of Gould's Belt[5] and the Local arm. Olano found that the Sun became attached to the local system about 500 Myrs ago and argued that this could be connected to a Galactic theory of terrestrial catastrophism.

A study by King et al. (2003), based on Hipparcos parallaxes and additional information, found that the likely age of the UMa MG was 500±100 Myr, thus Sirius itself could not be a member because of its significantly younger age.

7.4. Conclusions

The arguments of a very dynamical and fragmented LISM cause one to wonder whether the line of sight to Sirius could have been significantly changed in the last 2,000 years. Frisch (1995) suggested that the Sun entered the *local fluff* only a few 10^3 yrs ago. The possible small structure features

---

[5] A discrete system of primarily young stars that forms part of the Orion (local) arm of the Milky Way.

in the direction of nearby stars have a column density too low to produce significant foreground extinction, if one assumes the usual MW gas-to-dust ratio and ISM extinction properties.

Despite the similarity in motion, it does not appear that Sirius could be a co-eval member in the UMa moving group, since it is too hot and young. Accepting the proposal of Olano (2001) that the Sun, which is much older than the UMa MG, became attached to the moving group, one wonders whether Sirius could also have been attached in a similar way or could have been formed after most of the stars in the group were formed.

In the considerations of the neighborhood of Sirius one seemingly minor item stands out as unusual. This is the possibility of a very hot 10^5 K envelope around the Sirius system proposed by Bertin et al. (1995) for which there is no good explanation today.

Chapter 8
The perspective of stellar structure

In the previous chapters I explored the plethora of observational data now
available for the Sirius system, as well as for the space region in which Sirius
is now located or was located in the recent past. This chapter describes
models that explain separately the structure of Sirius A and of its WD
companion.

8.1. Upper main-sequence stars and Sirius A

The stars consist, most of the time, of equilibrium configurations of very
hot gas. On short time scales, measured in millions of years for many
objects, their dimensions or energy output do not change appreciably. This
is because the shape of any single star is determined by the balance of central forces whose common center is the center of the star. On the one hand,
self-gravity tends to pull the star together into this physical center. On the
other hand, gas pressure (because of the high temperature), and radiation
pressure from the photons created in the central region of the stars, tend
to expand the star. The steady-state condition when these forces balance
is called "hydrostatic equilibrium" and is the basis of understanding the
physics of stars. This balance of forces is maintained throughout most of a
star's life; on the main-sequence the source of energy that keeps the material hot and the photons flowing from the stellar interior to the rest of the
Universe is the fusion of Hydrogen nuclei to Helium.

The reactions that transform four protons into a ^{4}He nucleus can
proceed by different ways with the same end result: $4p \rightarrow {}^4\text{He} + 2e^+ + 2\nu$.
The transformation induced by nuclear fusion has one final outcome; the
production of energy in the interior of the star at the expense of the free-
proton content of the star. This energy generation arises because of the
difference between four times the rest mass of a proton to the rest mass of
one Helium nucleus.

A particularly simple description, from a physical point of view, of the
issues connected with stellar structure and stellar evolution (see Chapter 9)
was put forward by Prialnik (2000). Her description, drastically summarized
below and in the next chapter, can serve as a general introduction to the

subject. In this chapter I add more specific references to papers relevant to the Sirius system.

The energy production through the burning of Hydrogen can take place by fusing together four protons through intermediate light nuclei until a Helium nucleus is produced in the p–p chain, or by adding protons to the heavier Carbon, Nitrogen, and Oxygen nuclei in the CNO cycle. Each complete reaction of converting four protons into a Helium nucleus produces about 26 MeV, but the p–p and the CNO cycles have a different temperature dependence. The rates of energy production go as T^4 for the p–p chain and as T^{16} for the CNO cycle. At similar densities of stellar material, the energy yield from the CNO process is much more sensitive to small temperature changes.

Stars that are not on the main-sequence can produce energy by fusing three Helium nuclei into ^{12}C via the triple-alpha reaction, by fusing two Carbon nuclei into Magnesium, Sodium, Neon, or Oxygen, and by fusing two Oxygen nuclei into Sulfur, Phosphorus, Silicon, or Magnesium. Silicon, at a sufficiently high temperature, can lead to the production of the "stable" nuclei Iron, Cobalt, and Nickel. These nuclei are called *stable* because they have the highest value of the nuclear binding energy per nucleon among all nuclei; therefore, it is not possible to extract energy by their inclusion into heavier nuclei (fusion) or by dissociating them into lighter ones (fission). As a general rule, each fusion reaction involving heavier nuclei requires a higher threshold temperature to initiate and produces less energy per reaction that the fusion of lighter nuclei. The last stage, from Silicon to the Iron group nuclei, requires a temperature higher than 3×10^9 K so that the fusing nuclei can overcome the Coulomb repulsion.

Calculating the stellar structure implies solving a set of differential equations that model different dependencies among physical parameters describing the stellar interior. These yield the pressure, the temperature, the density, the chemical composition, etc., as a function of the distance from the center of the star. An important ingredient of stellar models is the way the energy is transported from its locus of production to the surface of the star. This depends on the opacity of the stellar material to the passage of radiation. Sometimes, the energy transport takes place by convection, that is, by the physical motion of hot blobs of stellar matter higher into the stellar envelope, with subsequent cooling by the emission of radiation, which can then escape from locations with smaller opacity.

If the internal parameters of a star do not change significantly, the energy-production rate remains fairly constant and the star is in thermal equilibrium. In addition, the self-gravity that tries to shrink the star is balanced by diverse sources of internal pressure that aim to expand it, and the star finds itself also in a state of hydrostatic equilibrium. A remarkable feature of such a state is that of *secular stability*; if the internal energy production decreases for some reason the internal pressure decreases as well, and thus the self-gravity pulls the stellar material inward to the center and both the density and temperature increase. The rate of nuclear fusion reactions depends strongly on the temperature; the star's interior will fuse light nuclei at a faster rate and the energy production rate will increase. A star that is in secular stable equilibrium will, therefore, function as its own thermostat.

Once a star has exhausted the Hydrogen in its core, it stops being a main-sequence object. Hydrogen fusion may continue in a shell surrounding the core, and the core will grow in mass because Helium nuclei will constantly be "raining down" on it from the burning shell. During shell burning the core contracts and the envelope (the part of the star external to the burning shell) expands. The star becomes at this stage a "red giant", with a much larger diameter but with a cooler effective temperature than it had while it was on the main-sequence. Further development stages depend on the mass of the star. Massive objects may develop advanced stages of burning, as described above.

A first calculation of the internal temperature and density distribution for Sirius A was published by Marshak & Blanch (1946). This was done in order to predict the luminosity of the star using new opacity tables available at that time. Marshak & Blanch assumed a pure-Hydrogen star (zero Helium content) and obtained luminosities of order $(1–8) \times 10^3$ $L_\odot$. In order to reconcile the model results with the actual luminosity of Sirius (38.9 $L_\odot$), they concluded that Sirius A must contain significant amounts of Helium.

Model atmospheres for Sirius A were constructed by Bell & Dreiling (1981). This study was triggered by new observational results that became available, such as the mass (Harris et al. 1963), and the radius (from a parallax determination by Jenkins 1963 and using the angular diameter measured by Hanbury Brown et al. 1974). The results from the models were compared with published and newly-acquired observations. These yielded log g=4.3±0.1 from a comparison of the predicted and measured Balmer line profiles. The effective temperature was found to be ~10,200±250 K and the abundance ratios [Fe/H] and [Ti/H] were calculated to be +0.6 and +0.5, respectively. The fluxes in the UV range predicted by the models were compared with those measured by Code & Meade (1976) using the OAO-2 satellite. Sirius was consistently found to be fainter in the UV than predicted by the models; this faintness could not be reproduced with metal-rich atmospheres.

A determination of radii for 12 main-sequence stars of spectral types A0 to G2 was done by Shallis & Blackwell (1980). They used accurate infrared photometry combined with parallax values from the literature, and compared the results with values determined from eclipsing binaries. Sirius was one of the objects they studied and the radius they found was 1.77 $R_\odot$. This is similar to, but slightly larger than the value found from intensity interferometer measurements (1.66 $R_\odot$, Code et al. 1976).

Both Sirius A and Sirius B appear to be X-ray sources, as indicated first by *Einstein* observation and confirmed later by *Chandra* imaging. Some have argued that Sirius A, which is a $\sim10^4$ K main-sequence star, could not be the source of significant X-ray production, and that its apparent detection is a result of a small amount of UV radiation penetrating the blocking filter of the X-ray detector. An explanation for the possible X-ray emission from Sirius A as a result of acoustic flux heating of a stellar corona was proposed by Fontaine et al. (1981). They calculated the expected acoustic flux by modifying a code developed for white dwarfs so that it would be applicable to main-sequence stars. The results showed that there exists a wide range of parameters where heating the Sirius atmosphere by acoustic fluxes could produce the required X-rays to account for the observed emission.

8.2. White dwarfs and Sirius B

One of the late stages of stellar evolution is a configuration that is not supported against collapse by radiation pressure, because nuclear reactions do not produce significant amounts of energy in the interior. In order for a main-sequence star to reach such a stage it must generally pass through at least one complete stage of fusion-produced energy, the main-sequence one, which ends when the core of the star is composed of Helium nuclei.

Stars of relatively high-mass manage to reach further fusion stages, such as that when Helium is burning to Carbon, Carbon is burning to Magnesium or Silicon, and even reach stages when the core is composed solely of Iron. Nuclear fusion reactions beyond Iron are endothermic, as explained above, thus a star cannot produce energy by fusing Iron group or heavier nuclei.

From a theoretical point of view, the HR diagram is a plot of the loci of stellar models on the plane defined by the stellar luminosity $L/L_\odot$ and the effective temperature T_{eff}, and both axes are plotted as logarithmic quantities. An observational HR diagram, where absolute magnitudes are plotted against color or spectral type, can always be transformed to the axes of the theoretical HR diagram.

The dominant location in the HR diagram, which is populated by most stars, is the main-sequence. The lower left-hand corner of the HR diagram harbors two kinds of stars, subdwarfs and white dwarfs. The subdwarf stars are those stars in the HR diagram located between the main-sequence and the WDs. There may be, in principle, two kinds of subdwarfs: those closely related to main-sequence stars, which are young, and those closely related to WDs, which are old.

The subdwarf stars are approximately one to two magnitudes below the main-sequence and more or less parallel to it. The WDs form another sequence approximately parallel to the main-sequence, but more than two orders of magnitude (5 mag) fainter that main-sequence stars of the same effective temperatures. This is, as mentioned above, a consequence of their small dimensions. The WDs in the HR diagram plot (Figure 8.1) are the nine lowest points. The source of internal pressure, which keeps the WDs from further collapse, originates from the degenerate electrons.

It was already mentioned here that the fainter companion of the Sirius system is a white dwarf. Here the "white dwarf" term will be further explored, and the evolutionary link between the present-day white dwarf and its main-sequence progenitor will be clarified. White dwarfs were recognized as a different type of star when Adams (1914) discovered that the star 40 Eridani (Eri) B, known to be under-luminous, had the spectrum of an A star and white colors. While the fact that the star was underluminous was known before, astronomers thought that it was a red dwarf, a main-sequence star of small luminosity and low effective temperature. Adams' discovery revealed that there are hot stars of much smaller luminosity than their main-sequence counterparts. This could only be explained if the surface area of 40 Eri B was much smaller than that of a main sequence star with the same T_{eff}; this star was small but white, thus hot. The source of the generic name *white dwarf* (WD) is in a short note written by Luyten (1922), who identified such small, faint, but hot stars based on their large proper motions.

Originally, the astronomical community recognized only three types of WDs. The largest group was that of WDs showing Hydrogen absorption lines in their spectra. This group was labeled DA, where D indicated that this star was a dwarf and A was intended to remind one of the principal characteristic of spectral type A: deep and prominent Hydrogen Balmer absorption lines. About 8% of the WDs did not show Balmer absorptions but did exhibit Helium absorption lines; these were called DB white dwarfs. Finally, the third category were the DC white dwarfs, about 14% of the total, and they showed no absorption lines at all.

Greenstein (1960) refined this rough classification and recognized ten types of white dwarfs. In his classification one takes note of a star's temperature along with the elements present in its atmosphere. The "other" types of WDs proposed by Greenstein, apart from the DA type (those showing Hydrogen absorption lines), the DBs (with Helium, but not with Hydrogen), and DCs (with no lines), were DF (with ionized Calcium), DG (with

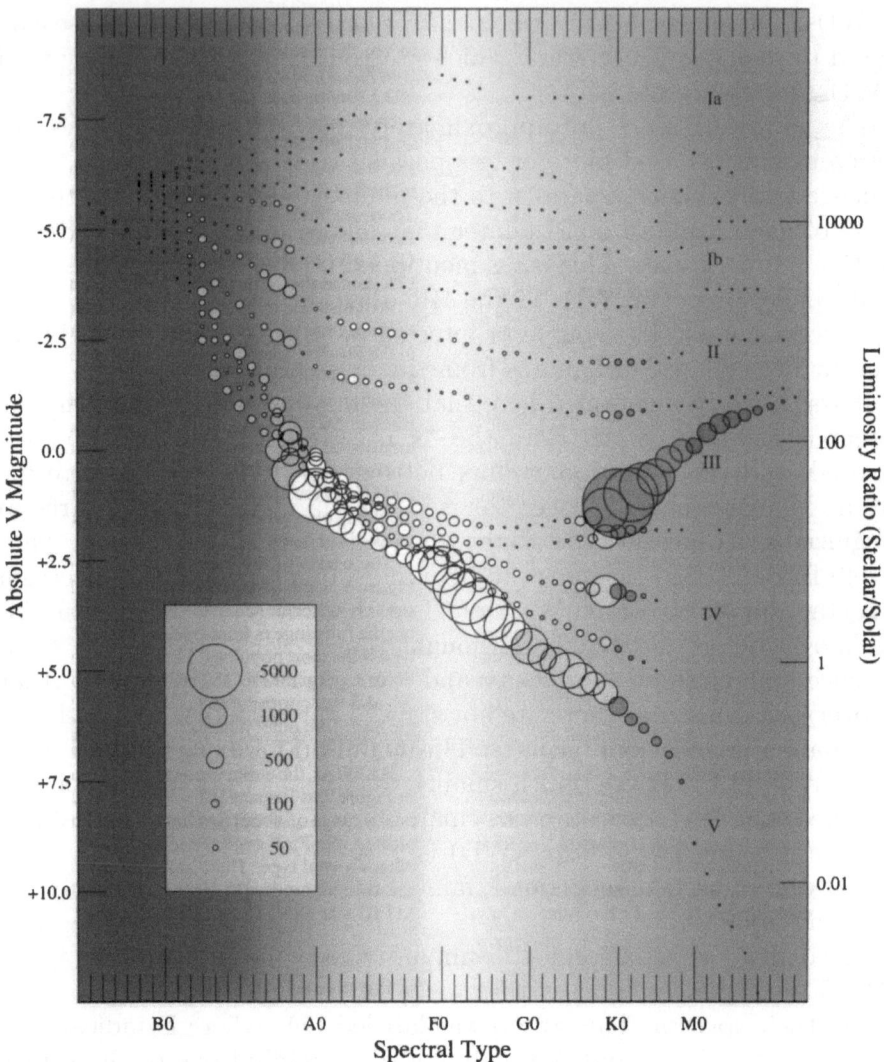

Figure 8.1. HR diagram based on stars with spectral classification by Houk (from Sowell et al. 2007). The spectral types are marked on the horizonatal axis and the luminosity classes are indicated to the right of the plot. The areas of the circles scale to the number of stars per spectral type and luminosity class. There are 113,286 stars plotted.

ionized Calcium and Iron), DM (with Ca II, weak Ca I, and TiO bands) and those of type C2 (or "λ4670 stars"). In addition, there were PG 1159 types which pulsate (very hot WDs with Carbon and Oxygen lines). Some WDs did not fit these types: e.g., LHS 1126, a WD as cool as the Sun that shows simultaneously Helium, molecular Hydrogen, and molecular Carbon features (Bergeron et al. 1994).

Sion et al. (1983) proposed yet another way to classify WDs. This simplified somewhat the complex symbols, which required Greenstein (1960) to add subscripts to indicate polarization, emission, and other visible characteristics of the spectrum. The newer classification scheme was based on only four symbols, with the first being D to symbolized degeneracy. Sion et al. (1983) proposed to add the second capital letter to represent the secondary spectroscopic features, if present in any part of the spectrum (e.g., Z, for a WD that shows also metallic lines in addition to its primary characteristic). The last part of the classification (not included in Table 8.1) is a digit that represents the effective temperature and is defined as $10 \times \frac{50,400}{T}$. These characteristics are clearly demonstrated in the *Atlas of Optical Spectra of White-Dwarf Stars* (Wesemael et al. 1993).

The spectral classification of white dwarfs, among all types of stars, was also discussed by Jaschek & Jaschek (1987) who adopted the scheme proposed by Sion et al. (1983). Astronomy recognizes today six main classes of white dwarfs, according to the dominant absorption lines in their (visible) spectra, which are described in Table 8.1.

WDs are special stars because they do not harbor Hydrogen fusion, nor any kind of light element fusion, in their interiors. One wonders, therefore, what keeps these stars from collapsing to their centers, since the radiation pressure originating from nuclear fusion is absent. Sir Ralph H. Fowler[1] (1926) proposed that white dwarf stars are supported against collapse by degenerate electron pressure. Subramanyan Chandrasekhar[2] (1939) worked out in detail the proposal by Fowler, introducing relativistic corrections and showing that there is a limit to the mass that can be supported by the degenerate electron pressure. Chandrasekhar was awarded the Nobel Prize for Physics in 1983 for his contribution to the understanding of stellar evolution.

TABLE 8.1. Recognized types of white dwarfs and their characteristics

Spectral type	Characteristics
DA	Only Balmer lines; no He or metals
DB	He I lines; no Balmer or metals
DC	Continuous spectrum; no deep lines (<5% of the continuum)
DO	He II strong; Balmer or He I present
DZ	No Balmer or He; metal lines only
DQ	Carbon features, either atomic or molecular

[1]Ralph H. Fowler, OBE (1889–1944), British physicist and astrophysicist.
[2]American astrophysicist (1910–1995) born in India.

Degenerate electrons are a peculiar state of matter in which the concentration of electrons becomes large compared with the quantum concentration (Phillips 1994). In quantum mechanical terms, the ensemble of free electrons in the star is at its lowest possible energy state (ground state), but this does not mean that the electrons have all the same energy. An electron is characterized by a set of quantum numbers that indicate its energy, its spin, etc. According to the theory of electrons, and to the Fermi-Dirac statistic that electrons follow, each system of electrons cannot have two particles with all the quantum numbers exactly identical. Because of this exclusion rule, there is no way to condense any further a concentration of electrons that is in the ground state. The electron pressure is, therefore, the internal factor that balances gravity and keeps the white dwarf star from collapsing further.

The energy output of any celestial object is measured through its luminosity L. The specific emissivity σ, the rate of energy emission per unit area of the star's surface, is determined by the effective temperature T_{eff}, the temperature of a star's outer layer from which the photons we observe are emitted. Most white dwarfs are good black body approximations thus for them $\sigma = \kappa \, T^4$. Among two stars with the same T_{eff}, the one with a smaller luminosity must have a smaller surface area, thus a smaller radius. Sirius B emits only 3×10^{-3} of the energy output the Sun produces. For the same T_{eff} as the Sun it should have a radius smaller by $\sqrt{3 \times 10^{-3}} = 0.05$ times the solar radius, i.e., some 35,000 km. This is what makes such an object a "dwarf", in comparison with a main sequence star. If it is hot, its color would tend to be seen as white, because most of its energy, and indeed the peak of its spectral energy distribution (SED), would be shifted far from the visible into the ultraviolet or even into the soft X-ray domain.

Chandrasekhar explained that white dwarfs are "highly underluminous" objects and as a suitable example he took Sirius B. With a mass approximately that of the Sun, it has only 3% of the solar luminosity. The physical difference, he mentioned immediately, is the "exceedingly high" value of the mean density. In white dwarfs, this can go up to 10^6 or 10^7 gr cm^{-3} whereas in the Sun it is ~ 1 gr cm^{-3}. The results of Chandrasekhar's calculations are a relationship between the mass and the radius of cold stellar bodies. Figure 8.2 shows this mass-radius relation with the axes' labels written in Tamil, Chandrasekhar's native language.

Electrons become degenerate at densities greater than 10^6 gr cm^{-3}=10^9 kg m^3. Such values are reached in the cores of some stars at the completion of the core Hydrogen burning, or are the rule in white dwarf stars. Under such conditions the internal pressure in a star is weakly dependent on temperature, depends mainly on the matter (electron) density, and can

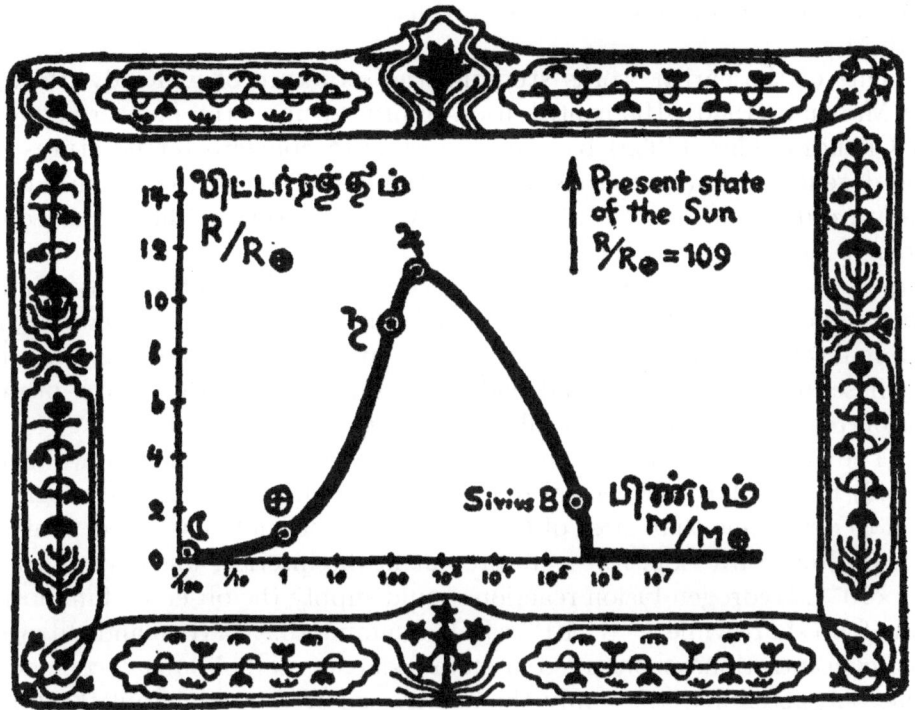

Figure 8.2. The mass-radius relation derived by Chandrasekhar and Kothari, as shown by Gamow (1940). The axes' labels in the diagram are in Tamil, Chandrasekhar's language. The left part of the curve plots Solar System bodies: the Moon, the Earth, Saturn, and Jupiter (from left to right up to the peak).

be described as $P \propto \rho^{5/3}$ for the non-relativistic case, or as $P \propto \rho^{4/3}$ for the relativistic case. These relations are called *equations of state* for the degenerate stellar matter.

Chandrasekhar's derivation for a configuration of a zero-temperature gas, for the non-relativistic case (low mass stars), gives the following mass-radius relationship:

$$R/R_\odot = 0.012\,(M/M_\odot)^{-1/3}\,(\mu_e/2)^{5/3}, \tag{8.1}$$

where μ_e is the mean atomic weight per electron. In the relativistic case, there is a maximal mass that can be supported by electron pressure:

$$M_{Ch} = 5.83\,(\mu_e/2)^{-2}\,M_\odot. \tag{8.2}$$

This is the well-known Chandrasekhar limit, beyond which the degenerate configuration collapses into a neutron star or into a black hole. For $\mu_e = 2$, which is the case for a Helium WD (two electrons per each Helium nucleus), the Chandrasekhar limit is 1.46 $M_\odot$, while for μ_e=2.15 (Iron WD) it is only 1.26 $M_\odot$.

One of the earlier works studying the structure of white dwarfs was by Marshak (1940). His contribution improved upon the pioneering work of Chandrasekhar (1939) by considering finite non-zero temperatures for the degenerate configurations through the derivation of the internal temperature distribution of white dwarfs. An application of the stellar equations to the case of Sirius B, considering the observable parameters of mass, radius, and temperature, yielded the necessary conclusion that only minute amounts of Hydrogen could be present in the interior. This is because the internal temperature of the WD should be at least ten million degrees and, in these high-temperature and high-density conditions, the proton-proton fusion reaction would have resulted in a large energy output.

The integration of the stellar structure equations allowed Marshak (1940) to calculate the central temperature of Sirius B: 15.2×10^6 degrees, and to determine that most of the star is ~isothermal. The upper limit to the Hydrogen fraction, if it were present in the interior, was derived to be 2.2×10^{-8}. Hydrogen fusion reactions could supply the observed luminosity only for a short while ($\sim 8 \times 10^5$ years), assuming that Carbon and Nitrogen had concentrations as observed in main-sequence stars. Marshak concluded that one possibility to supply the energy required to maintain the luminosity could be the release of gravitational energy. This seemed to him to be plausible, because of the small radii of white dwarf stars that implied a very strong gravitational potentials. This, in addition to the simple cooling of the WD; Marshak estimated that at the present luminosity of Sirius B its internal heat would suffice to keep it shining for 3×10^8 years.

Mestel (1952) discussed the energy sources of white dwarfs and concluded that no effective internal sources could exist since the time of the WD birth. Mestel found that the basic structure of a WD is of a degenerate core which is highly conducting, surrounded by a thin envelope of non-degenerate matter. This outer layer has a relatively high opacity and maintains the luminosity at a low level. In connection with possible external sources of energy, Mestel considered accretion of Hydrogen by a WD with possible nuclear fusion by either the proton-proton or the CNO cycle. This was done specifically for Sirius B, in order to explain a possible discrepancy in the mass-radius relation for this star. The discrepancy proved to be due to a mis-estimate of the radius, not requiring the assumption of accretion onto α CMa B.

Schatzman (1958) summarized the status of knowledge of WDs and of the dense matter making up their interiors. He plotted the mass-radius relation for WDs for Hydrogen and for Helium WDs and showed that the experimental points lined up with the Helium WD relation. He also discussed the energy production in WDs and concluded that, if Hydrogen fusion takes place at all, it could provide only a very small fraction of the radiated energy. Energy production by stellar contraction was, however, deemed possible.

A further step forward in understanding WDs was made by Schwarzschild (1958), where the question of stellar structure and evolution was dealt with from a perspective of two decades after Chandrasekhar's contribution. Schwarzschild described various types of stars according to their location on the Herzsprung-Russell (HR) diagram.

The definitive contribution to the structure of white dwarfs was made by Hamada & Salpeter (1961), based on a more accurate equation of state. They calculated theoretical models of stars at zero temperature, for compositions of ^{4}He, ^{12}C, ^{24}Mg, ^{28}Si, ^{32}S, and ^{56}Fe. These nuclei are all debris of nuclear fusion that take place at different densities and temperatures in cores of stars. The important result of Hamada & Salpeter, which is still widely used 40 years after its publication, is a relation between the mass and the radius of each such configuration. This is shown in Figure 8.3.

The white dwarf models of Hamada & Salpeter (1961) differ in their radii from those calculated by Chandrasekhar mainly at the lower end of central densities, ρ_c. The deviations peak at about 20% less than the Chandrasekhar

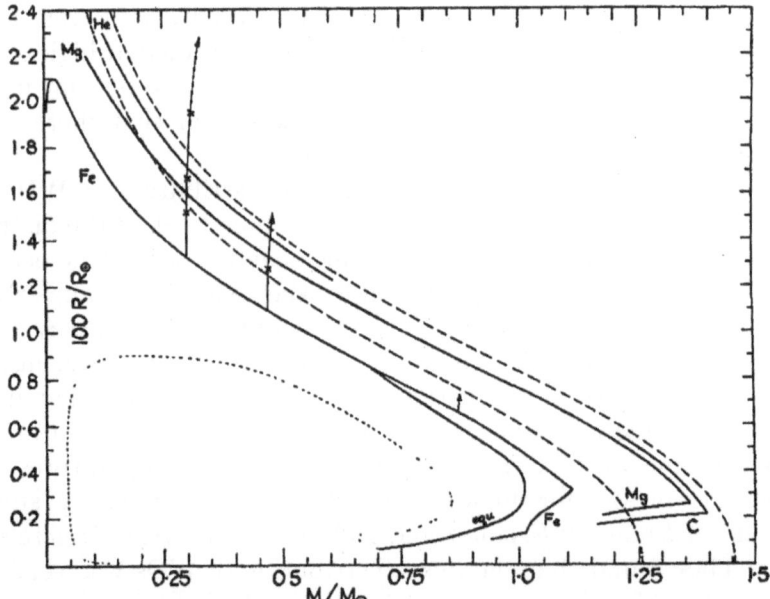

Figure 8.3. The original Hamada & Salpeter (1961) relation between the mass (*horizontal axis*) and the radius (*vertical axis*) of white dwarfs of different compositions. The almost-vertical arrows indicate the location of WDs with Hydrogen envelopes. The dashed curves represent neutron star models; the dotted ones are models by Chandrasekhar.

values for a 0.1 $M_\odot$ star with $\rho_c \sim 10^5$ gr cm^{-3}. In addition, there are different possible highest masses for WDs, depending on their composition. Instead of 1.46 $M_\odot$, as found by Chandrasekhar (1939) for the maximal mass of a Helium WD, Hamada & Salpeter calculated 1.40 $M_\odot$ for a pure ^{12}C object. Thin Hydrogen envelopes are possible, and their presence increases the radius of a compact star without changing significant its mass.

In the context of the general population of white dwarfs, Sirius B is outstanding. This is because of its high mass, 1.05 $M_\odot$ (Holberg et al. 1998) or 1.02±0.02 $M_\odot$ (Barstow et al. 2005), when compared with the observed mean masses of white dwarfs. Weidemann & Koester (1984) analyzed a sample of 70 DA white dwarfs from the general population with multichannel spectrophotometry from Greenstein (1982), and compared the observations to models. They concluded that the mean mass of a WD was 0.62 $M_\odot$ with a dispersion of 0.13 $M_\odot$. McMahan (1989) used a similar method but a larger sample and found a mean mass of 0.546 $M_\odot$ and a dispersion of 0.19 $M_\odot$. This is virtually identical to the determination by Bragaglia et al. (1995) for a sample of 45 southern sky WDs.

The average mass of a white dwarf was also estimated at 0.562±0.137 $M_\odot$ by Bergeron et al. (1992). In particular, no evidence was found for WDs with masses higher than $\sim$0.9 $M_\odot$ among single stars (Liebert 1980). The masses found later by Bergeron et al. (1995) for 43 WDs were slightly higher than typical ($\sim$0.6 $M_\odot$), but not very different. Reid (1996) found a median mass of 0.581 $M_\odot$ for a sample of 53 WDs studied at high resolution with the Keck 10-m telescope. The consensus seemed to be that most WD masses are significantly below 1 $M_\odot$, yet it was clear that there are WDs much more massive than these typical values.

In recent years, some claimed a bimodal distribution of WD masses. Vennes (1999) argued that, in addition to the main peak of the distribution at 0.57±0.02 $M_\odot$, there is at least one much smaller secondary peak at 1.2 $M_\odot$ (see his Figure 8). One explanation for the secondary peak is that such ultra-heavy WD were formed by the merging of two Carbon-Oxygen WDs. Wickramasinghe & Ferrario (2000) argued that such WD merging may be linked to the presence of magnetic fields in the WDs, thus that magnetic WDs may more likely be the progenitors of type Ia supernovae[3].

Napiwotzki et al. (1999) re-determined effective temperatures, surface gravities, and masses for 46 white dwarfs selected from EUV surveys of the *Extreme Ultraviolet Explorer* and *ROSAT Wide Field Camera* satellite missions. They found that the mass distribution of WDs was very sharply peaked at 0.589 $M_\odot$; the underlying distribution had to be extremely peaked at this value to yield the shape found, considering the experimental errors.

---

[3]Type Ia supernovae are thought to be the result of the collapse of a WD that, due to accretion or other reasons, reaches a mass larger than the Chandrasekhar limit.

The discussion presented above on WD masses emphasizes that α CMa B is outstanding by being significantly more massive that the typical WD. At the high-end of the masses there are a few WDs even more massive than Sirius B. Trimble (1999) listed such examples: G35-26 with 1.2 $M_\odot$, PG 1648+441 and PG 0136+251 both a bit over 1.2 $M_\odot$, and a few WDs ranging up to 1.43±0.06 $M_\odot$. The last value "scratches" the theoretical upper limit for the maximal mass of a white dwarf as set by Chandrasekhar.

Greenstein et al. (1977) presented results from very high resolution spectroscopy of 14 WDs of type DA that showed sharp absorption cores in the centers of their broad Balmer limes. These sharp cores allowed Greenstein et al. to conclude that their projected rotational velocities must be lower than 40 km s^{-1}. The slow rotation of the WDs implies that during the mass-loss stage (see Chapter 9 angular momentum was lost together with the additional mass.

In a series of papers in the 1960s and 1970s, Greenstein and Trimble studied the gravitational redshifts of white dwarfs (Greenstein & Trimble 1967; Trimble & Greenstein 1972). Their first paper published radial velocities for 92 WDs and the second had another 74. Combining all radial velocities, after accounting for the space velocities of the stars, yielded a mean redshift of +54±7 km s^{-1} (N=102 stars). For Sirius B itself, Greenstein et al. (1971) measured a redshift of +89±16 km s^{-1}, as already mentioned above, pointing out this WD as exceptional.

Using the Hamada & Salpeter (1961) mass-radius relation for Carbon white dwarfs (see below), Koester (1979) derived a radius of 7.76×10^{-3} $R_\odot$ for Sirius B. This, together with the observed V magnitude and model fluxes, yielded T_{eff}=27,400±1,000 K.

The X-ray emission from Sirius B was explained also by a class of theoretical models developed by Dziembowski & Gesicki (1983). The X-rays could originate from acoustic flux generated not in the subphotospheric convective zone of the WD's atmosphere, as generally assumed, but by short-period oscillations in the layered envelope, if the hydrogen-rich layer is sufficiently shallow.

Theijll & Shipman (1986) determined the radius and rotational velocity of Sirius B from the effective temperature set by Holberg et al. (1984) and from theoretical work by Ostriker & Hartwick (1968). The determination of the radius is based on the theoretical relation between magnetic field, radius, and angular momentum. Assuming no magnetic field and a homogeneous rotation for Sirius B, Theijll & Shipman found Sirius B to be slightly above the Hamada & Salpeter (1961) mass-radius relationship. In order for it to fit the relation, they had to assume 1σ shifts in the magnitude, mass, and parallax. The rotational velocity of the WD, in this case, should lie between 0 and 600 km s^{-1}.

Koester (1987) presented measurements of the gravitational redshift in nine DA WDs and showed that these follow the Hamada-Salpeter zero temperature models for Carbon WDs. The difference with other studies was that Koester was able to account for the space motions of the WDs in his sample by deriving this from the systemic velocity, when the WD was a member of a wide binary with a normal primary star, or when it belonged to a common proper-motion pair.

The mass-radius relationship for white dwarfs was tested by Provencal et al. (1998) using *Hipparcos* results. The motivation was to improve WD masses, which were obtained through a comparison of experimental spectra with model atmospheres. Such comparisons yield surface gravities (log g) and effective temperatures (T_{eff}) by fitting line profiles. Sirius B was one of their primary targets, for which they found that the parallax measured by *Hipparcos* fitted to 1σ the ground-based measurement of Gatewood & Gatewood (1978). The revised mass they found, 1.000 ± 0.016 M$_\odot$, and the slightly different radius, $(7.4\pm0.7) \times 10^{-3}$ R$_\odot$, result mainly from the adoption of a different effective temperature. This value, $24,700\pm300$ K, comes from the study by Kidder et al. (1991) of broad-band (U–B vs. B–V) colors for a sample of WDs, where the measured colors were over-plotted on a grid of log g and T_{eff}.

Provencal et al. (1998) used gravitational redshifts of WDs in common-proper-motion pairs, and surface gravities and effective temperatures of field WDs, all objects having accurate *Hipparcos* parallaxes, to determine these accurate radii. Their results are shown in Figure 8.4. Sirius B is the rightmost point in the diagram; it does not seem to fit a Hamada-Salpeter Carbon-Oxygen configuration. However, the experimental point fits very well the mass-radius relation of Wood (1995) for a Carbon core, thick Hydrogen atmosphere WD with $T_{eff} \approx 30,000$ K.

Koester et al. (1998) performed high-resolution spectroscopy of the Hα cores in a sample of 28 DA white dwarfs, attempting to detect the rotation of this kind of objects. Their results are compatible with no rotation, with upper limits of $v \sin i < 15$ km s^{-1}. Given that in these stars there was also no detection of any Zeeman splitting, Koester et al. were also able to limit the magnetic field strength; this could have been at most a few tens of Kilograms. The result implies that an efficient transfer of angular momentum took place during the evolution to the WD stage, probably during the strong mass-loss phase. Mass-loss during post-main-sequence stages will be briefly discussed in the following chapter.

Note that the most recent values for the mass and radius of Sirius B (Barstow et al. 2005) fit admirably well the Hamada-Salpeter relation. This was demonstrated above in Figure 6.5.

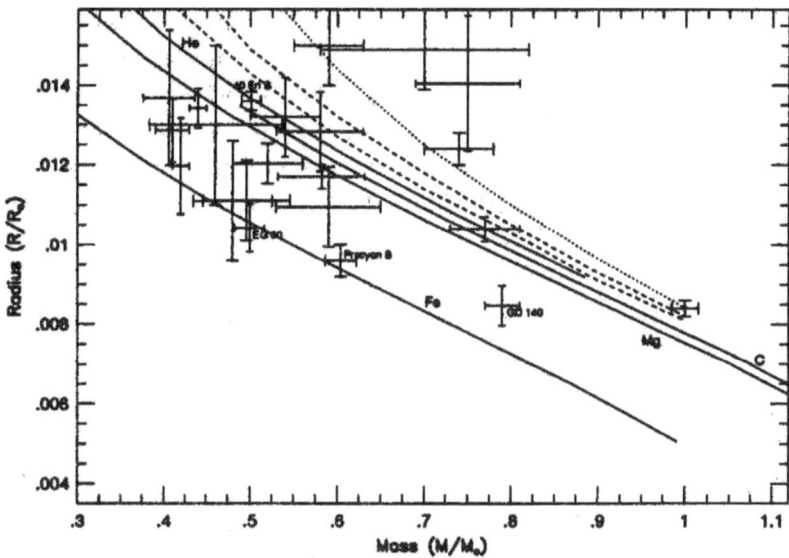

Figure 8.4. Testing the mass-radius relationship of Hamada & Salpeter (1961) with accurate measurements of white dwarfs (from Provencal et al. 1998). Sirius B is the rightmost point in this diagram, with the highest mass among the plotted WDs.

8.3. Conclusions

We saw here that, while Sirius A appears to fit reasonably well most of the expected properties of an A1 main-sequence star, its WD companion is outstanding in its high mass, when compared with the general WD population. The slow spin of α CMa A in comparison with the majority of A stars remains unexplained and in many cases is taken as a fluke. The spin of the WD was not reported by the best available observation, that of Barstow et al. (2005). It, and eventually a measurement of the magnetic field, might influence ideas about the system's past evolution.

Chapter 9
The perspective of stellar evolution

Given that the Sirius system consists of a main-sequence A0 star and a very massive white dwarf, it is of interest to discuss its evolution. As there are two stars in the Sirius system, I will describe first, in general terms, the evolution of stars more massive than the Sun following their main-sequence stage. Then I will describe the evolution of the brighter component A, followed by that of the fainter one. This will be done mainly by considering each object as evolving in isolation. I will then describe briefly, but from a general point of view, the evolution of stars in a binary system. The description of the evolutionary process for the specific Sirius system will be presented in Chapter 10.

The previously-mentioned work by Schwarzschild (1958) is not only a good reference for a classical treatise on the stellar structure, but also presents a comprehensive treatment of the stellar evolution. Since for the case of Sirius discussed here one is not primarily interested in the mode the stars reached the main-sequence, I shall skip this development stage and will explain the evolution only since the more massive star in the system left the main-sequence. The development of the stars, in particular after the main-sequence stage, can be described as evolutionary paths in a theoretical Herzsprung-Russell diagram, as shown in Figure 9.1.

The different stages indicated in Figure 9.1 are (following Iben 1967 and Prialnik 2000) 1–2 main-sequence, 2–3 overall contraction, 3–6 Hydrogen burning in shell, 6–7 red giant phase, 7–10 Helium core burning. The duration of each phase, for a few selected stellar models, is given in Table 9.1.

One clearly sees from Table 9.1 that each evolutionary time off the main-sequence is considerably shorter than the time a star resides on the main-sequence. The entry in the Table for the 2.25 $M_\odot$ model corresponds approximately to Sirius A. The model fitting Sirius A, with a mass of $\sim$2 $M_\odot$, should be between that for the 2.25 $M_\odot$ and the one for the 1.5 $M_\odot$ models. This does not account for a possible increase in the mass of Sirius A from its original main-sequence stage, due to the possible accretion of material from Sirius B while it was a red giant (see below). The progenitor of Sirius B should be one of the higher-up entries in the Table between 5 and 9 $M_\odot$.

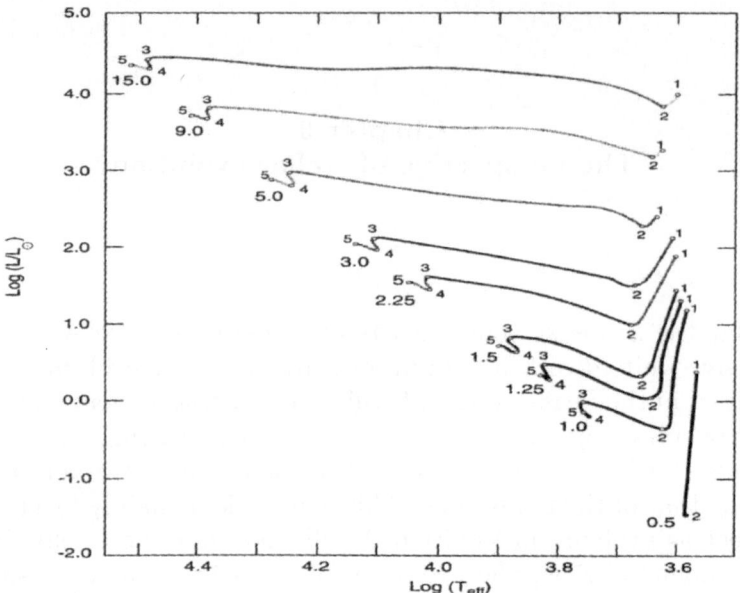

Figure 9.1.   Theoretical HR diagram showing the evolutionary paths off the main-sequence for metal-rich stars of different initial masses, after Iben (1967). Various development stages are numbered.

An important stage of development, after a star leaves the main-sequence, takes place when the core is composed of Carbon and Oxygen as a consequence of previous burning stages. Such stars, at a post main-sequence phase, become asymptotic giant branch (AGB) objects, an extension of the giant branch of the HR diagram to stars of even higher luminosity. These stars are on their way to become super-giants, exceedingly bloated stars with very high luminosity.

The structural characteristics of AGB stars, following Prialnik (2000), are:

1. Nuclear burning takes place in two shells, one above the other, the inner burning Helium and the outer burning Hydrogen,

TABLE 9.1.   Lifetimes of stars with different masses (years) for different stages of evolution

$M/M_\odot$	1–2	2–3	3–6	6–7	7–10
15	$\times 10^7$	2.3×10^5	7.6×10^4	7.2×10^5	8.45×10^5
9	2.1×10^7	6.1×10^5	3.07×10^5	4.9×10^5	3.55×10^6
5	6.5×10^7	2.2×10^6	2.64×10^6	6.1×10^6	1.09×10^7
2.25	4.8×10^8	1.6×10^7	8.8×10^7		
1.5	1.6×10^9	8.1×10^7	3.5×10^8	1.0×10^8	$>2 \times 10^8$

2. Luminosity is determined only by the core mass,
3. Intense stellar wind because of strong radiation pressure in the envelope.

The important consequences of these structural features are that the two-shell burning is unstable and is prone to sudden short and strong enhancements (thermal pulses). The strong stellar wind implies that, in this stage, the star exhibits a significant mass loss, ranging from 10^{-9} to 10^{-4} $M_\odot$ yr^{-1}. The mass loss can cause the dispersion of the stellar atmosphere into the interstellar medium, or can form a temporary concentration of mass around the mass-losing star: this in known as a planetary nebula (PN). Note that the ejected mass is not stationary; it is expanding steadily and eventually it dissolves in the ISM.

As a consequence of the mass loss, stars in the mass range of ~ 1 $M_\odot$ to ~ 9 $M_\odot$ are left with stellar cores made of Carbon and Oxygen that have masses from 0.6 to 1.1 $M_\odot$. These are quickly transformed into white dwarfs as the thin Hydrogen shell left on top of the core can no longer sustain nuclear burning (at a shell mass of 10^{-3}–10^{-4} $M_\odot$).

9.1. Evolution of a main-sequence A star: Sirius A

An early attempt to calculate the evolution of a star similar to Sirius A was reported by Henyey et al. (1955). They followed the development of a star with a mass on 2.291 $M_\odot$, with 74% Hydrogen (X value), 25% Helium (Y values), and 1% metals (Z value, by weight) from a late proto-stellar configuration to the main sequence and beyond. Their very early calculations showed that the star never reaches the triple-α regime, when three Helium nuclei fuse into a Carbon nucleus. After 6.5×10^8 years, its effective temperature is 4600 K.

Proper modelling of the structure and evolution of Sirius A requires good knowledge about its physical parameters. Hartmann et al. (1975) calculated stellar evolutionary tracks appropriate to Sirius A given the then-known observational parameters: the effective temperature measured by Code et al. (1976) and the mass of the star, determined using the parallax and the orbital parameters of Gatewood & Gatewood (1978). They attempted to fit the age of Sirius A assuming that the Sirius group, to which α CMa A was claimed to belong, is similar to an open cluster of co-eval stars. The age for the Sirius group could be derived as $\sim 3 \times 10^8$ years. With this calibration, Sirius appeared to be too hot for its color or spectral type. Hartmann et al. found that they could fit models to Sirius either if it is younger by a factor of ~ 2 than the Sirius group, or if they allowed it to have a substantially smaller metallicity than the other stars in the group (Z=0.015, vs. 0.03 for the Sirius group stars).

Evolutionary models must be tested against actual observations. A comparison relevant for the early-type stars was performed by Schönberner & Harmanec (1995). They found a good correspondence between the brightness values predicted by the theory of Schaller et al. (1992) and those measured for stars in binary systems, for which there exists an accurate determination of their physical parameters (T_{eff}, mass, etc.). They showed that many of the stars they checked fit well onto the zero-age main-sequence (ZAMS), including Sirius.

A special type of comparison is that of the age determination for the main-sequence stars performed by Lachaume et al. (1999). They determined ages for stars of spectral types B9 to K9 within 25 pc using five different methods: location in the HR diagram compared with theoretical isochrones, rotational velocity, strength of photospheric Ca II emission lines, stellar metallicity, and space velocity. The over-abundance of metals in Sirius prevented an age determination using a metallicity indicator. Neither was any other method successful in determining an age for Sirius and, in their final table (Table A.6), Lachaume et al. left the entry for Sirius blank.

The metallic-line (Am) stars were first described by Titus & Morgan (1940) as stars showing a particular spectroscopic anomaly. Conti (1965) stated that Sirius shows abundance anomalies like those in Am stars. In conjunction with rotational properties, he showed in his Figure 2 the distribution of $v \sin i$ against the B–V color for bright northern A and F stars. From this plot, it is clear that all the Am stars are slow rotators, with $v \sin i \leq 70$ km s^{-1}. For the specific case of Sirius, Conti proposed that the abundance anomalies could have been produced by contamination from Sirius B, before it became a WD.

Conti (1970) proposed a different definition of Am stars from that of Titus & Morgan (1940), such that these would be those stars

> that have an apparent under-abundance of Ca (and/or Sc) and an apparent overabundance of the Fe group and heavier elements.

One of the characteristics of Am stars, Conti mentioned, is their low rotational velocity. This has already been mentioned above, when rotation was discussed in the chapter dealing with modern observations of Sirius (Chapter 5). While normal A stars have $v \sin i \approx 150$ km s^{-1}, Am stars have $v \sin i \approx 55$ km s^{-1}. The question, therefore, arises whether the Am phenomenon is a cause or a result of the low rotational velocity. As there are some normal A stars that are slow rotators, it seems that this is a necessary condition but not a sufficient one for a star to be Am. Conti concluded, therefore, that the abundance anomalies are real, and not some form of stellar atmospheric effects. The effects may be caused by some physical separation of the elements in a star's atmosphere.

The evolution of Am and Fm stars was studied by Richier et al. (2000). Their stellar models included radiation-driven diffusion of metals and radiative acceleration. The assumption was that there is convection overshooting, which homogenizes the interfaces between the Hydrogen, Helium, and Iron-peak zones, and produced surface abundance patterns that closely resemble those of Am and Fm stars. For Sirius A, 16 measured or upper-limit abundances were compared with the predictions. Of these, 12 were well-reproduced by the model, three were not so well reproduced, and one (S) was disregarded as it was deemed to be an unreliable determination.

Richier et al. (2000) studied Sirius A in particular because it is one of the best–observed Am stars. It seems that the surface abundance of many of the elements for which observational values exist (He, Li, O, Na, Mg, Si, P, Ca, Ti, Cr, Fe, and Ni) can be reproduced accurately by stellar models of 2.3–2.5 $M_\odot$, which have the mixing depth at $10^{-5.3}$–$10^{-4.6}$ $M_\odot$ at an age of 100 Myr. However, the calculations of Richier et al. failed to reproduce the surface abundance of C, N, and Mn. In their conclusions, Richier et al. (2000) mentioned that the discrepancy between the models and the observations, in the case of Sirius, lie mostly in the CNO abundances. Those would be largely reconciled, if the CNO would have been processed by the CNO cycle. They also proposed, as did others before, that Sirius B, in an earlier stage of evolution, could have transferred processed material to Sirius A. The amount of transferred material should have been at least 10^{-5} $M_\odot$ to produce a significant effect. Note however that the abundance analysis of Lambert et al. (1982) seems to preclude this possibility, because the total abundance of Carbon, Nitrogen, and Oxygen is different from what would result from the CNO cycle.

Liebert et al. (2005) used the TYCHO stellar evolutionary code of Young & Arnett (2005) to fit the observed parameters of Sirius A. These were the radius (1.711±0.013 $R_\odot$; Kervella et al. 2003), the mass (2.02±0.03 $M_\odot$), and the luminosity: 25.4±1.3 $L_\odot$. They found a best-fit age of 237.5±12.5 Myr assuming solar composition for the primordial abundance and membership in the Sirius supercluster. Assuming Z=0.5 $Z_\odot$ they found a good fit for an age of 375±19 Myr. Note that they did not assume mass transfer from the Sirius B progenitor to the A component took place.

9.2. Evolution of a white dwarf: Sirius B

Following the cessation of nuclear reactions in the thin atmosphere left on top of the core of a fairly massive star after most of the envelope is ejected, a white dwarf has only passive sources of heat to use as sources of radiation. The main source is the thermal energy stored in the ions. At much later stages, the crystallization of the ions provides a secondary heat source. The evolution of WDs is, therefore, solely described in the HR diagram by cooling curves. An important question, relevant to the description of the evolution of the Sirius system, is the mass of the progenitor of Sirius B while it was on the main-sequence, since this sets the size of the red giant; this has implications for the evolution of a binary system. This has been dealt with specifically or generically in a number of papers, as will be described below. The conclusion seems to be that the main sequence progenitor of Sirius B was a $\sim$ 7 $M_\odot$ star, though a recent publication (Ferrario et al. 2005) lowers this to $\sim$5 $M_\odot$.

Auer & Woolf (1965) explained the presence of WDs in the Hyades, as well as of a suspected white dwarf in the Pleiades, by the fact that stars with masses greater than 2.5 $M_\odot$, even as massive as 7 $M_\odot$, can become WDs. They argued that this is because mass loss can occur as a result of rotational shedding at the star's equator if the star rotates as a solid body. Mass loss at this stage may prevent the star's demise in a supernova explosion that would leave behind a neutron star or a black hole, and may instead leave a WD remnant.

A special attempt to determine the masses of the precursor stars of cluster white dwarfs was made by Jones (1970). This was based on the determination of the age of the cluster, from the main-sequence turnoff point, and on surveys for white dwarfs in these clusters. The conclusion of the study by Jones was that stars heavier than 1.86 (−0.12, +0.48) $M_\odot$ on the main-sequence will become WDs on the brighter branch of the WD color-magnitude diagram.

The main-sequence turnoff point is the location, in the HR diagram, where stars belonging to a single-age population (such as stars in a cluster that formed virtually simultaneously) begin their evolution toward the red giant or super-giant stages. At the turnoff location, the points representing cluster stars do not scatter anymore around the main-sequence but flex off toward the giant branch. Since the location of a star on the main-sequence is only a function of its mass, the turnoff point marks the location of the most massive stars that are still on the main-sequence, still burning Hydrogen in their cores. Therefore, this turnoff point is an age indicator for single-age (co-eval) stellar populations.

The question of mass loss from a star on its way to become a WD was also treated by Weidemann (1977). He suggested that, depending on the angular momentum of a pre-WD star, the outcome of the evolution may be a WD even for a precursor as massive as 10 $M_\odot$. The mass loss after the main-sequence phase would reduce the mass so much that a supernova explosion may not occur. Weidemann showed in his Figure 1 (reproduced below as Figure 9.2) the dependence of the final mass of a star on its initial (main-sequence) mass. The experimental points are plotted in the Figure together with different theoretical relations. The plot indicates that, for

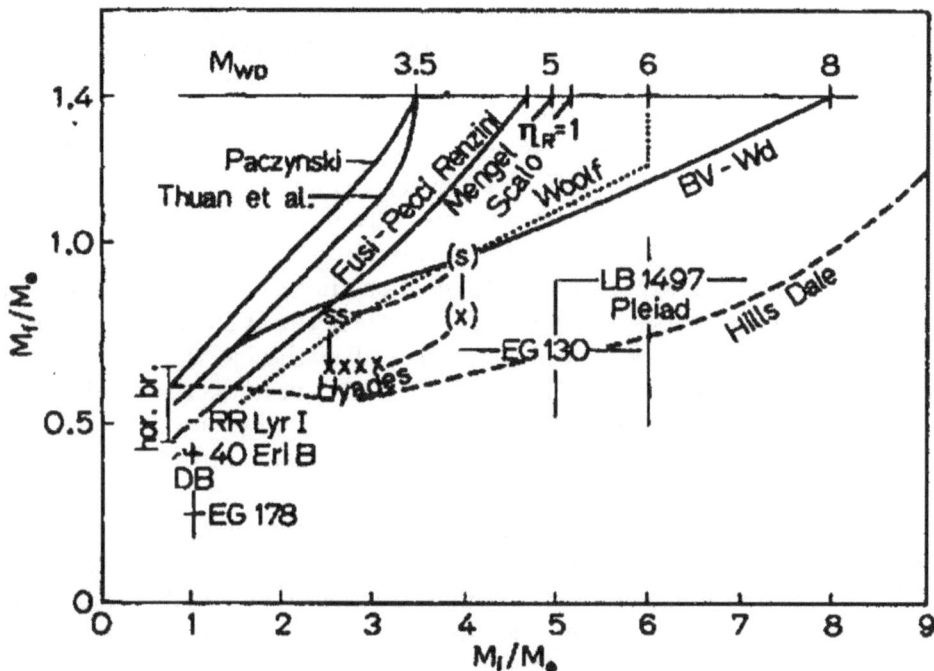

Figure 9.2.   Initial (*horizontal axis*) vs. final (*vertical axis*) mass for single white dwarfs (from Weidemann 1977). Mass estimates for a few objects are compared with different theoretical predictions. The best-fitting relation seems to be that of Hills & Dale (1973).

high-mass WDs, the relation proposed by Hills & Dale (1973) seems to fit best, despite the arguments against it put forward by Fusi-Pecci & Renzini (1976).

Weidemann (1977) argued that one should adopt an upper mass for a WD precursor of 5 $M_\odot$, but considered this value to be a statistical average, relevant for calculations of galactic evolution models, with higher or lower values possible. He argued that the results should not be relevant for Sirius B, since the relation should be valid only for single WDs whereas α CMa B is a member of a binary system, and also hinted that mass transfer between the components could have taken place. However, the present distance between the components is such that there could hardly be an evolutionary influence of one star upon the other. If high initial angular momentum, through effective outward transport, favors high mass-loss rates, then the more massive a WD precursor star would be, the more massive the resultant WD should be. Weidmann cautioned finally that the predictions of Hills & Dale (1973) may be biased towards lower final masses; thus the upper masses for a WD progenitor may even be higher than calculated by them.

The general question on the rapidity of evolution of WDs was studied by D'Antona & Mazzitelli (1987) in connection with the internal structure of WDs. The external appearance of a WD, for instance the presence or absence of Hydrogen in the atmosphere, is related to the thickness of the remnant Hydrogen and Helium layers.

D'Antona & Mazzitelli (1987) described schematically a WD as a set of concentric entities: a Carbon-Oxygen core, a Helium mantle, and a Hydrogen envelope (present in most cases). This outer envelope is apparently absent in all non-DA white dwarfs. Most of the mass of the WD resides in its core; this also provides the thermal reservoir of the WD. Heat is released from the core after the WD formation, i.e., after the ejection of the stellar envelope, and when the WD crystallizes; the questions of whether the result is a disordered process, whether Carbon and Oxygen are not mixed in the crystallizing core, and whether the Oxygen separates and sinks to the center, were not yet solved when they wrote the paper.

The importance of the Helium mantle in the evolutionary process, with specific application to Sirius B, lies in it being a *buffer zone* between the outer Hydrogen atmosphere and the inner Carbon. It is possible that, during the sedimentation of the heavier nuclei, Carbon nuclei may encounter Hydrogen (protons) with a resultant exoenergetic thermonuclear reaction:

$$p + {}^{12}C \rightarrow {}^{13}N + \gamma. \tag{9.1}$$

This Hydrogen flash, induced by the Carbon diffusion, may have been common in white dwarfs and could have been related to the historical records of Sirius being red (cf. Section 4.1).

Joss et al. (1987) studied the evolution of giant stars, with particular emphasis on the determination of the core mass. In particular, they considered the core mass-radius relationship as a test of stellar evolution models. Sirius B is one of the special cases they treated in the context of Roche lobe overflow and rapid transition to a white dwarf stage in some 10^3 years. They combined a number of theoretically-determined relations into parametric equations, and found, from a 1986 private communication by Eggelton, that it is possible to derive a luminosity-core mass relation of the form:

$$L = \frac{10^{5.3}}{1 + 10^{0.4}\,\mu^4 + 10^{0.5}\,\mu^5}\, L_\odot, \tag{9.2}$$

where $\mu = M_c/M_\odot$ is the core mass parameter. They then determined the radius of a giant star as a function of the core mass parameter from evolutionary tracks calculated by Paczyński (1970):

$$R \approx \frac{3.7 \times 10^3\,\mu^4}{1 + \mu^3 + 1.75\,\mu^4}\, R_\odot. \tag{9.3}$$

These relations enabled Joss et al. (1987) to discuss the evolution of a WD progenitor that is a member of a binary system with specific application to the Sirius system. They concluded that it was very likely the system was an interacting binary at the time the progenitor of Sirius B was a red giant.

Mazzitelli & D'Antona (1987) discussed the structure of WDs, in particular the faster cooling of non-DA WDs. For these, the atmospheric opacity is reduced in comparison with white dwarfs with Hydrogen atmospheres. The mixing of the Hydrogen atmosphere into the underlying Helium reduces the opacity and enhances the cooling; a Hydrogen WD (of type DA) effectively switches cooling tracks and finds itself onto the Helium cooling track. During their evolution, the WDs show stratification of the different nuclei; the core of a Carbon-Oxygen WD would consist mainly of Oxygen, with a Carbon-rich mantle. Downward percolation of Hydrogen from a DA WD atmosphere through the Helium mantle, and the encounter of protons with an upward rising plume of Carbon, could cause localized and intense Hydrogen burning; this is the Hydrogen flash already mentioned.

The cooling of white dwarfs was summarized in a thorough review paper by D'Antona & Mazzitelli (1990). The process is determined by E, the energy content of a WD, and by $\frac{dE}{dt}$, the rate at which the energy is radiated away:

$$L(t) = -\frac{dE}{dt} \qquad (9.4)$$

and the terminology is driven by the recognition that the largest contribution to the radiated energy comes from the reduction in the thermal energy content of the white dwarf. A more complete relation accounts also for nuclear energy (L_{nuc}), if any hydrogen burning takes place through percolation to deeper levels, for gravitational energy release (L_g), and for neutrino cooling (L_ν).

D'Antona & Mazzitelli (1990) began by considering the evolution of the WD precursor star. This, they took to imply, is a star that already departed the asymptotic giant branch, with a thin Hydrogen envelope, which burns Hydrogen via the CNO cycle. A WD, on the other hand, does not burn Hydrogen. The WD progenitor star should have been heavier than $M_{min} \approx 0.8$–1.0 $M_\odot$, depending on the age of the galactic disk and on the WD chemical composition, but cannot be more than $M_{up} \sim 9$ $M_\odot$. Some WDs are stellar remnants following Helium burning, thus they are composed of Carbon and Oxygen. A few WD progenitors, with mass close to M_{up}, may have experienced Carbon burning, ending with a core of ~ 1.2 $M_\odot$ composed of Oxygen, Neon, and Magnesium. This process is more likely to occur in close binary systems, where mass loss from the more massive star that evolves faster to the lighter star may preclude core collapse.

The more massive the progenitor star is while on the main sequence, the higher the percentage of Carbon relative to Oxygen in the core. WDs originating from a 1–3 $M_\odot$ star may contain 70–80% Oxygen; if their progenitor was a 5 $M_\odot$ star, Oxygen would only be $\sim$50% of the core mass. The strong gravitational field of the compact WD produces a sorting of the nuclei by sedimentation, so that the top layer is composed of almost-pure Hydrogen (if any) on top of an almost-pure Helium layer, which lies on top of the Carbon mantle. Between the layers there is no clean separation; a tail of Hydrogen penetrates the Helium layer from the top and a tail of Carbon sneaks up through the Helium by diffusion, driven by the large concentration gradients of the different nuclei. The stratified atmosphere acts like a set of blankets on top of the isothermal core, determining its cooling rate. The schematic structure of the outer layers of two WDs, one of 0.6 $M_\odot$ and one of 0.84 $M_\odot$ is shown in Figure 9.3.

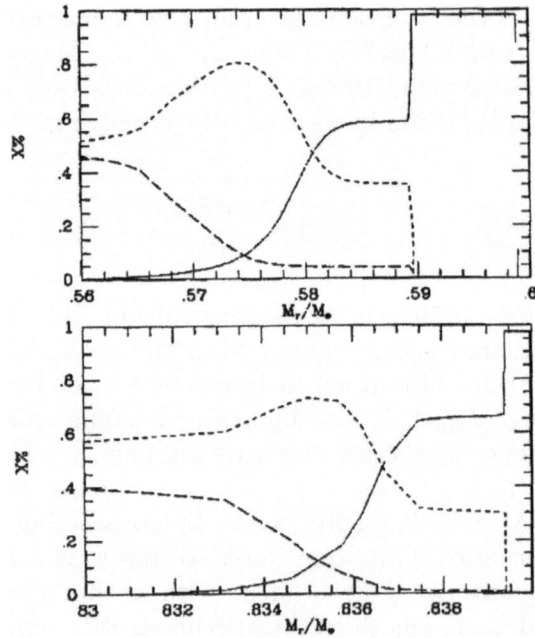

Figure 9.3. Atmospheres of two WDs (Figure 1 from D'Antona & Mazzitelli 1990). The top panel shows the stratification of a remnant from a 1.0 $M_\odot$ star now an 0.6 $M_\odot$ WD; the bottom represents the stratification in an 0.84 $M_\odot$ remnant from a 3.0 $M_\odot$ star. The dotted curves show the distribution of ^{4}He, the short-dashed ones represent ^{12}C, and the long-dashed ones the distribution of ^{16}O.

At the beginning of the WD phase, the radius of the star can be up to twice that of a zero-temperature fully-degenerate configuration (D'Antona & Mazzitelli 1990). Young WDs can, therefore, heat up slightly by contracting their outer non-degenerate layers. Another heat source can materialize if the Helium buffer layer between the Hydrogen atmosphere and the Carbon interior is sufficiently thin, as already mentioned above, and allows Hydrogen plumes to reach the Carbon interior. Hydrogen burning, in this case, is explosive and produces a *Hydrogen flash* (self-induced nova; Iben & McDonald 1986).

An interesting source of energy at late stages of WD development is the release of latent heat during the crystallization of the WD interior, already mentioned above. Electrostatic interactions cause the nuclei to come together, the star behaving like a liquid, then to arrange themselves as a body-centered-cubic (BCC) lattice. This re-arrangement releases latent heat since it is a first-order phase transition[1], and the released energy can increase the cooling time by ~30% for a one solar mass Carbon WD (Lamb & Van Horn 1975).

A number of approximate formulae have been put forward linking initial pre-WD masses (M_i) with final WD masses (M_c). For example, Vassiliadis & Wood (1994) included the metallicity and proposed:

$$M_f/M_\odot = 0.473 + 0.084\,(M_i/M_\odot) - 0.058\,\log(Z/Z_\odot). \qquad (9.5)$$

With the measured mass of Sirius B, and for an almost-Solar metallicity, the progenitor of Sirius B should have been a ~7 $M_\odot$ star.

The cooling time of a WD to reach a certain luminosity was given by Kawaler (1996, quoted in Kwok 2000, p. 171) as:

$$t_{cool} = 9.41 \times 10^6\,(A/12)^{-1}\,(\mu_e/2)^{4/3}\,\mu^{-2/7}\,(M/M_\odot)^{5/7}\,(L/L_\odot)^{-5/7}\,yr \qquad (9.6)$$

Here A is the atomic weight of the ions in the degenerate core and μ is the mean molecular weight of the envelope. For a C-O WD, the average is $A \approx 14$.

While the behavior of the internal parts of a WD seems to be fairly well understood, D'Antona & Mazzitelli (1990) pointed out the difficult situation regarding the energy transfer through the WD atmosphere, where the matter is in a partly degenerate, partly ionized state. The internal parts of the star are highly conductive, because of the good thermal conductivity of degenerate electrons (several orders of magnitude better than metallic Silver or Copper at laboratory conditions). However, the outer layers show a complicated behavior, with convection and radiation both playing a role in the outward transport of energy. Convection, in particular, may be extremely important in facilitating the Hydrogen flash already mentioned above, since it may bring Hydrogen-rich material into the inner parts of the WD.

---

[1] A first-order phase transition involves latent heat. During such a transition a system either absorbs or releases instantaneously a fixed and large amount of energy.

A computation relevant to Sirius B was performed by Winget et al. (1987) for pure Carbon WD models with masses from 0.4 $M_\odot$ to 1.0 $M_\odot$. They considered the luminosity distribution of the white dwarfs in the solar neighborhood as a constraint on the age of the galactic disk (and through that, on the age of the Universe). Specifically, Winget et al. argued that the sudden drop in the space density of low-luminosity WDs, at $\log(L/L_\odot) \approx -4.5$ is caused by the lack of suitable WDs that did not have sufficient time to cool to such low brightness. In order to estimate cooling times, they calculated models of WDs. The one model relevant for Sirius B is that for a one solar mass Carbon WD. A perusal of their Table 1 shows that such a WD reaches a luminosity similar to that of Sirius B ($\sim$0.01 $L_\odot$) after 274 Myrs.

Weidemann (1990) reviewed the status of WD masses and evolution, including also the WD progenitors. In this context, he mentioned the relation $M_f(M_i)$ derived by Weidemann & Koester (1983): the relation is approximately flat for M_i lower than 1–3 $M_\odot$ and bends upward for higher masses to $M_f \approx 8$ $M_\odot$.

The evolution of WDs in the context of Galactic evolution was studied by Salaris et al. (1997). They analyzed the influence of the nuclear reaction $^{12}C(\alpha, \gamma)^{16}O$ on stellar nucleosynthesis and found that, for the best choice of convection and reaction rates, Oxygen will be enhanced in the central regions. This would affect the cooling of the WD and would increase its life by 1.5 Gyr. Their calculations for the evolution of a star that leaves a 1.0 $M_\odot$ C-O white dwarf from a 7 $M_\odot$ progenitor showed that the lifetime of the WD progenitor is 4.5×10^7 year.

Another study of the rather massive (3–9 $M_\odot$) stars, relevant for understanding the origin of type Ia supernovae, was performed by Umeda et al. (1999). Type Ia SNe are produced by the collapse of a very massive WD, with $M \approx 1.37$–1.38 $M_\odot$. If mass accretion takes place, such a massive WD may explode because of Carbon ignition in its central region. The intention of Umeda et al. was to understand the evolution of the intermediate-mass stars and, in particular, of the Carbon-Oxygen cores.

Without going into the technical details of the Umeda et al. (1999) models, it is sufficient to say that they found a Helium-burning shell around a central mass of 1.1 $M_\odot$ in a 7 $M_\odot$ stellar model with Z=0.001. This indicates that the Carbon-Oxygen core of this model will ultimately form a WD heavier than 1.1 $M_\odot$ following the envelope ejection. In fact, this may even result in an Oxygen-Neon-Magnesium WD. For a higher metallicity (Z=0.02) and the same initial mass, the core was found to be a C-O concentration of 0.95 $M_\odot$. The lower the mass of the WD progenitor, the lower the mass of the resulting WD. Figure 6 of Umeda et al. (reproduced here as Figure 9.4) shows the dependence of the C-O mass on the ZAMS mass of the progenitor, as a function of metallicity.

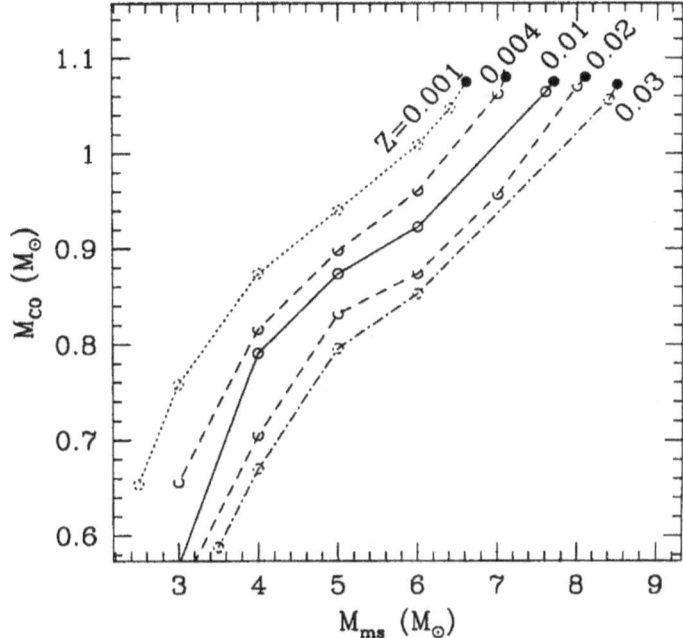

Figure 9.4. Initial vs. final mass for single white dwarfs (from Umeda et al. 1999). The different curves represent different metallicities of the progenitor star. The uppermost points of each track indicate that off-center Carbon ignition occurs in these models, thus they would be the highest masses for Carbon–Oxygen cores.

The figure shows that in order to complete the evolution with a C-O core of 1.05 $M_\odot$, a progenitor of 6.5–8.2 $M_\odot$ is required (the heavier masses required to have higher metallicities). Similar results for the highest possible mass of a WD progenitor, which will yield a high-mass WD, were obtained by Bono et al. (2000).

Kwok (2000) gathered a number of progenitor mass estimates for final WD masses in a figure, shown here as Figure 9.5 and fitted the distribution with a smooth curve. Note the sharp increase in the slope of the relation at the high-mass end; this is a result of the specific fourth-degree polynomial fit used by Kwok. The points themselves do not appear to support such a sharp increase and, in fact, a ~1.2 $M_\odot$WD could result from the evolution of a ~8 $M_\odot$ star. Sirius B could be the remnant of a ~ 6.5–8 $M_\odot$ star.

A reconsideration of the initial-to-final masses of WDs was done by Ferrario et al. (2005) using stars in open clusters and "anchoring" their derived relation to the point (M_i=1.1±0.1 $M_\odot$; M_f=0.55±0.01 $M_\odot$) in order to reproduce the peak in the mass distribution of field WDs. Using a linear relation fit, the precursor mass of Sirius B is obtained as ~5.5 $M_\odot$. With this relation, and allowing curvature, they also fitted a sixth-degree polynomial

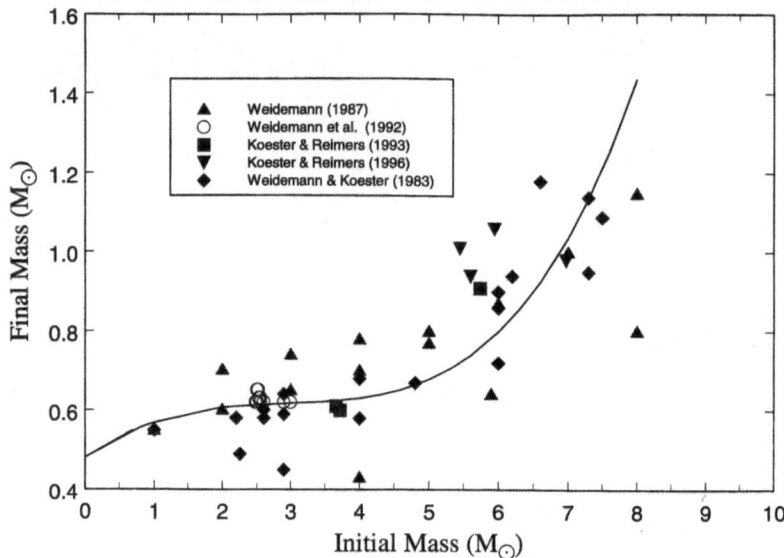

Figure 9.5.   Initial vs. final mass for single white dwarfs (from Kwok 2000).

to the points. With it, the progenitor mass should have been a ∼6.5 M$_\odot$ main sequence star.

The issue of rotation and magnetic fields in WDs was discussed by Trimble (1999). She mentioned the general assumption that both angular momentum and magnetic flux are conserved throughout the evolutionary process leading to a WD configuration. WDs are, typically, slow rotators compared with their break-up rotation periods of tens of seconds. The measured $v \sin i$ values of 5–15 km s^{-1} in DAs that show sharp cores in their Balmer lines (Koester et al. 1998) imply periods of one or more hours. In magnetic WDs periods as long as 80-years have been measured.

Liebert et al. (2005) modelled Sirius B, as already mentioned above. They used the recent values for the mass, 1.00±0.02 M$_\odot$, while adopting T$_{eff}$=25,000±200 K and log(g)=8.528±0.05 (Barstow et al. 2005), and derived the age of Sirius B as a WD from its cooling time. They used the WD cooling sequence of Fontaine et al. (2001) and fitted T$_{cool}$=123.6±10 Myr for a WD model with half its core mass of Carbon and half of Oxygen. The TYCHO stellar evolution code they used indicates that the progenitor mass should have been between 4.78 and 5.065 M$_\odot$. A slightly larger mass, $5.123^{+0.28}_{-0.23}$ M$_\odot$, could be obtained using the Girardi et al. (2002) tables. These are the "lightest" mass estimates published for the progenitor of Sirius B.

9.3. Binary star evolution: Sirius as population representative

Many nearby stars do not exist as single stars, but rather as members of multiple-star systems. The most frequent are binary systems, two stars that orbit each other. van de Kamp (1969) considered a sample of 59 stars nearer than 5.2 pc from the Sun. Among these, 22 are in binary systems, six are members of two triplets of stars, and among the 30 objects that appear single (including the Sun), six seem to have unseen companions. Approximately one-half of the nearby stars are single; the other half are members of multiple systems of stars. One should not, therefore, wonder that the existence as a binary system may have influenced the evolution of each of the two stars in the Sirius system.

The double stars have long been recognized as important keys in the understanding of stellar evolution. van de Kamp (1961) mentioned that 75% of the O and B stars in associations are double or multiple stars. As these are high-mass stars, their lives end in supernova explosions as explained above, and it is possible that during such a violent event the binary becomes disrupted with one of the stars being ejected from the system with a high velocity. This proposition was first put forward by Blaauw (1961).

Eggen (1986) considered the case of binary stars, specifically that of wide binaries. He showed that there is an apparent bias against very wide binaries in catalogues of binary stars, with under-representation of systems having separations of $\sim$0.1 pc and larger. Eggen advocated the definition of a binary as two stars that show parallelism of motion in all three vector components of their velocities. Under this definition, the Sirius system would have to be considered a close binary since the orbital motion and the joint space motion are easily detectable with modern observational means.

Note that there is no rigorous definition of what a "close binary" is. An operational definition could be that this is a system of two stars that are sufficiently close together that they influence each other's evolution. In practical terms, this may imply that the orbital period of the system is shorter than about 1 year (Harpaz 1995) and under such conditions tidal forces cause significant deformation of the components.

One special ingredient in calculations dealing with stellar evolution in a binary system, particularly for close binary stars, is the concept of the Roche[2] geometry. This subject is intimately related to the concept of Lagrangian points and equi-potential surfaces. These topics will be discussed briefly below.

In a binary star system, the gravitational potential assumes a somewhat complicated dependence on distance. This is because each member of the binary system affects material close to the binary. In addition, one has to

---

[2]Edouard Roche (1820–1883), French mathematician who studied celestial mechanics.

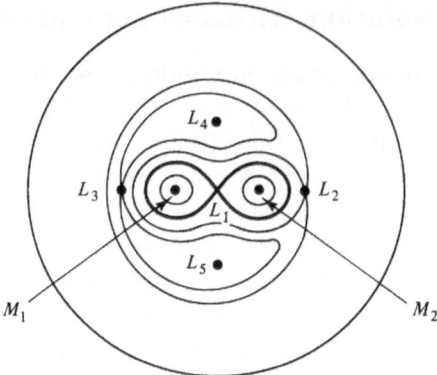

Figure 9.6. Equi-potential surfaces and Lagrangian points in a binary star system. The heavy contour joining the envelopes of the two stars is the cross-section of the Roche lobe along the orbital plane.

account for rotation due to the orbital motion of the two components and of any other material belonging to the system (that would be orbiting the barycenter). In the rotating system, in the plane of rotation of the two stars, there are five special points called *Lagrangian*[3] *points* and labeled L_n, with $n = 1 \rightarrow 5$, where small masses can be in equilibrium because the gravitational forces from the two stars are balanced by the centrifugal force due to the rotation of the entire system.

Three of the Lagrangian points are co-linear with the two stars. The inner Lagrangian point, L_1, is between the two stars, closer to the heavier member of the pair. L_2 is external to the lighter star and L_3 is external to the heavier star. The points L_4 and L_5 are on a line perpendicular to the central axis on which the first three Lagrangian points lie. The inner Lagrangian point L_1 is the location where two equi-potential surfaces, one around one of the stars and the other around the second, touch. This point allows the easiest location for mass transfer between the two stars, as will be made clear below. Figure 9.6 shows the equi-potential surfaces and the five Lagrangian points of a binary system.

The unique equi-potential surfaces that just enclose one or the other of the stars and touch at the L_1 point are called the *Roche lobes*. A matter element within one of the Roche lobes experiences a force in the direction of the stellar component enclosed by the lobe. If, during the stellar evolution, one of the components were to fill its lobe because of evolutionary effects, such as expansion during the red-giant phase, at least some of the matter near the star would pass through L_1 and fall onto the other component.

[3]Named after the French mathematician Joseph Louis Lagrange (1736–1813), born in Italy, who studied this configuration of masses.

This process is called *Roche lobe overflow*[4] and is responsible for the transfer of matter between components of a binary system.

Although Roche lobes do not have a spherical, or an easily describable form, it is possible to define a "Roche lobe radius" that is the radius of a sphere that has the same volume as that of the lobe. This radius can be calculated, in terms of the mass of the stars and their separation, from the relation (Iben & Tutukov 1984):

$$R_{i,L} \approx 0.52 \left(\frac{M_i}{M_{tot}}\right)^{0.44} A = F \times A, \tag{9.7}$$

where A is the distance between the stars and M_{tot} is the sum of the two stellar masses. Various terms have been derived for the multiplier factor F of the distance between the binary components. The formula approximated by Eggleton (1983) has a wide applicability:

$$F = \frac{0.49 \, q^{2/3}}{0.6 \, q^{2/3} + ln \, (1 + q^{1/3})}, \tag{9.8}$$

where q is the binary mass ratio.

The issue of mass transfer in binary systems was also raised by D'Antona & Mazzitelli (1978) in conjunction with the explanation of the X-ray emission from white dwarfs. They explained this as being produced in a $2 \times 10^{16} \leq M_H \leq 10^{17}$ gr (10^{-17}–2×10^{-16} M$_\odot$) corona of the WD. This amount of Hydrogen should have been accreted by Sirius B from the stellar wind of its companion. This implies a mass loss rate from Sirius A of $3 \times 10^{-13} \leq \dot{M} \leq 2 \times 10^{-12}$ M$_\odot$yr^{-1}, which can be sustained by an A-type star.

A thin, low-mass Hydrogen envelope accreted on a white dwarf may yield a very high luminosity in the EUV range (Shara et al. 1978). These calculations were done for an 0.8 M$_\odot$Carbon-Oxygen WD and showed that for an envelope of mass $(2.58$–$5) \times 10^{-5}$ M$_\odot$a thermal nuclear runaway can happen without mass ejection takeing place. More massive envelopes produce similar runaway burning and also eject part of the envelope. For $\sim$100 year the WD appears as a very luminous ($\sim 2 \times 10^4$ L$_\odot$) star and for the last 10 year of this period it appears as a soft X-ray source.

Iben (1991) showed that if the evolution of the binary includes mass transfer between the components after the Hydrogen has been exhausted in the core of the more massive star but Helium did not yet begin to burn there, the outcome could be WDs of different masses, from $\sim$0.5 M$_\odot$to the Chandrasekhar limit. Figure 9.7 shows the WDs produced in this situation.

---

[4]Strictly speaking, Roche lobe overflow means that the volume of the star exceeds the volume of the Roche lobe.

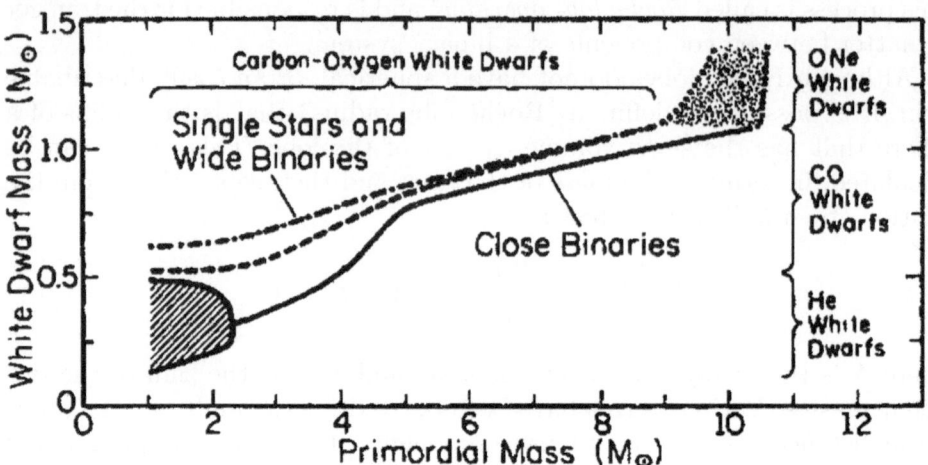

Figure 9.7.   Masses and compositions of white dwarfs as function of the mass of the progenitor star, in a Case B Roche lobe overflow (from Iben 1991). Sirius B should lie, according to its mass, near the top end of the Carbon–Oxygen WDs.

Sirius, in such a figure, would belong to the "Single Stars and Wide Binaries" category and the progenitor of Sirius B would then be a star of 8–9 $M_\odot$.

Binary stars that show magnetic fields may have their evolution influenced by magnetic braking. This may modify the rotation of one or both stars in the system, as explained by Li & Wickramasinghe (1998). Figure 9.8 shows schematically the magnetic field configuration in such a system.

The theory of Li & Wickramasinghe (1998) identifies as one of the more important parameters the angle between each magnetic field and the spin axis of the respective star. Each field is assumed to be a simple magnetic dipole, i.e., no magnetic multipole fields are present, as is the case for the Sun. The more inclined are the two dipole fields, the stronger the magnetic braking will be. In cataclysmic variables, for which the authors intend the analysis to be relevant, the inclinations must be less than 45°. This mechanism may have been relevant during the pre-WD phase of Sirius, and perhaps for a short time after the formation of the WD.

Hachisu et al. (1999) studied possible ways to produce Type Ia supernovae (SNe). They proposed that such situations of collapsing WDs may be encountered in symbiotic binaries[5] consisting of a WD and a low-mass red giant, where the source of mass to be accreted by the WD is the red giant wind. While the result they obtained about the frequency of Type Ia SNe is not relevant to the present discussion, their discussion dealing with the mode of increasing the event frequency is. In a nutshell, Hachisu et al.

[5]Long-period (200–1000 day orbit) interacting binaries having an ionized nebula in addition to the two stars.

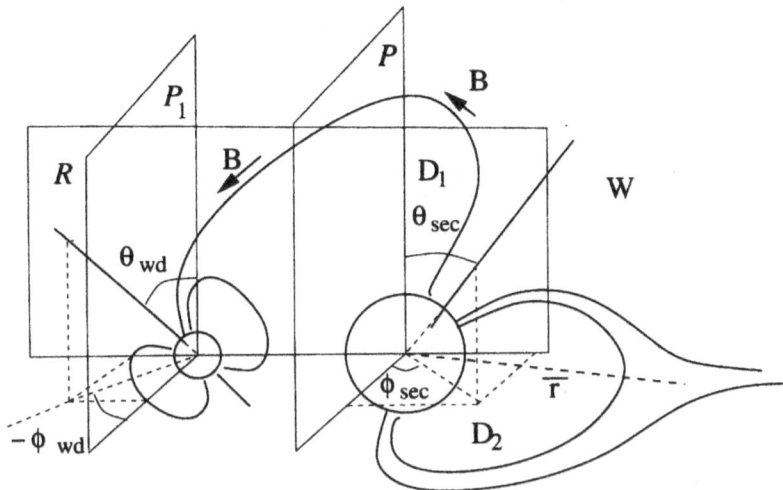

Figure 9.8.   Magnetic fields in binary stars. This is Figure 3 of Li & Wickramasinghe (1998) and illustrates the configuration of the magnetic field in a binary system. Region W represents the wind zone and region D is the "dead" zone, where the magnetic field lines of the two stars join up.

showed that a close binary consisting of a white dwarf and a red giant may form from a wide binary, even if the initial separation was as large as 40,000 $R_\odot$=187 a.u. The relevancy of this to the Sirius system is that the present separation of the components is smaller than this value, thus Sirius could fit this scenario requirement.

The key in bringing together the components of a wide binary is, according to Hachisu et al. (1999), a special condition of the superwind emitted by the more evolved component at the AGB stage. If the wind emitted by the AGB star has a velocity slower than the orbital velocity or equal to it, the transfer of angular momentum from the binary to the wind is efficient and the two stars approach each other. The full scenario proposed by Hachisu et al. may be relevant for the future development of the Sirius system, when the present Sirius A will reach the AGB stage. The first part, the reduction in the separation of the two stars, could have been relevant for the production of the present Sirius system, when mass loss from the WD progenitor may have taken place at such a low velocity as to bring the two stars closer than their initial separation.

9.4. Conclusions

We have seen how a star such as Sirius A might evolve when isolated, and how can a more massive but isolated star become a WD. Sirius A has an anomalous metal abundance that makes it an Am star; even the convection overshooting model of Richier et al. (2000) could not fully explain those.

Considered in isolation, Sirius B appears to have originated from a
5-8 $M_{\odot}$ main-sequence star. Many authors invoked mass transfer from the
progenitor of Sirius B to the A component to explain the enhanced metal
abundances.

The evolution of binary stars was also discussed briefly in general terms,
but special emphasis was put on explaining the Roche geometry and the
possibility of mass transfer in a binary, once the system leaves the main-sequence. The next chapter will present specific models designed in particular
to explain the Sirius system.

Chapter 10
Sirius revealed – a synthesis of the information

This chapter will summarize and combine the information presented in the previous sections of this book into a hopefully coherent picture of the Sirius system. I will show first that Sirius is not unique in its physical properties, but it is just one member in a population of binary stars. Following this, I will present some specific models calculated to explain the main characteristics of the α CMa system, including some that attempt to account for the historical redness. I will conclude with thoughts on the future on the Sirius system.

10.1. Sirius analogs

Are there other stellar systems similar to that of Sirius? At least some researchers think that Sirius is not unique and that other early-type main-sequence stars co-exist with white dwarfs in binary systems. IK Peg=HR 8210 is claimed to be such a system (Wonnacott et al. 1993). This object was first identified as a metallic-lined A star, but was subsequently recognized as a strong far-UV source from the all-sky survey with the Wide-Field Camera on *ROSAT*. Subsequent searches in the *Einstein* data set revealed a soft X-ray source at the position of IK Peg and an *IUE* observation showed the signature of a hot star, enhanced flux shortward of $\sim$1,650 Å, where the A star spectrum falls off rapidly.

Wonnacott et al. (1993) derived the mass of the white dwarf in the IK Peg system by first fitting an effective temperature from the *IUE* spectrum considering the constrains imposed by the X-ray observations ($T_{eff} = 33,000$ K), assuming a surface gravity of log $g = 8.0$, and then using the Hamada & Salpeter (1961) zero-temperature mass-radius relation. The mass of the white dwarf in the IK Peg system was determined to be 0.985 ± 0.03 M$_\odot$. The primary star in the system is an F2V or an A8m, and its estimated mass lies between 0.85 and 1.45 M$_\odot$. The system is a close binary with a period of 21.7 days and the binary orbit shows a small, non-zero eccentricity ($e \approx 0.027$). The short period implies a much closer separation of the components that in the α CMa system.

Wonnacott et al. (1993) gave other examples of Sirius-like systems. These are β Crateris (Al Sharasif), an A2IV or A1II star with a low-mass white dwarf companion in ~10 yr orbit (Burleigh et al. 2001), and V651 Monocerotis, an A5V star with a very hot (100,000 K) object in a 15.991-day orbit around it. This object is at the center the planetary nebula NGC 2346, which is ~6,000 years old. Although similar to Sirius, many of these systems have their components much closer to their primaries than the two α CMa stars.

To those one can add the peculiar system IN Coma, a possible triple system that lies at the center of the large and diffuse planetary nebula LoTr 5 (Brosch & Hoffman 1999). One of the components, presumably the star that produced the nebula, is an extremely hot subdwarf. Landsman et al. (1995) found that the binary IK Peg is composed of an A8V star and a massive WD with a mass greater than 1.17 M$_\odot$ (see also Parthasarathy et al. 2007, with M$_{WD}$ = 1.25 M$_\odot$). However, unlike the case of Sirius but similar to other examples above, this is also a close binary with a period of only 21.7 days.

Vennes et al. (1998) analyzed a sample of 13 hot DA white dwarfs, each paired with a luminous, unresolved companion. Their sample is dominated by non-WD companions of later types than in the Sirius system; only one companion is of type A1III (β Crateris, mentioned above). The observational information for the stars they studied was taken from the *EUVE* and *ROSAT* data sets, as well as from ground-based observations. In their analysis, they fitted model atmospheres to the FUV and EUV SEDs, as well as to the profiles of the Lyman α lines. The goal was to determine T$_{eff}$, log g, and log n_H (the column density of neutral Hydrogen in the WD's atmosphere). In addition, they also attempted to measure a radial velocity curve for each object that could be the result of orbital motion. The results of the study by Vennes et al. (1998) were that two high-mass WDs were identified: EUVE J0044+095=BD +08° 102 (~0.9 M$_\odot$) and EUVE J2126+193=HR 8210 (1.00–1.32 M$_\odot$). About half of the systems in their sample were found to have short orbital periods, of order days.

Gatti et al. (1998) studied the nucleus of the planetary nebula Abell 35. They used the Planetary Camera (PC) of the Wide-Field/Planetary Camera instrument on the *Hubble Space Telescope*, with a pixel size of 4.55×10^{-2} arcsec, to image the nuclear region of the nebula through the filters F300W, F336W, and F547M (central wavelengths 2,950, 3,350, and 5,465 Å respectively). The high angular resolution, and a careful comparison of the image of the nucleus with the theoretical point-spread function of the PC, demonstrated that the image of the nucleus is elongated. The spectral energy distribution of the nucleus showed the presence of two stars: one of spectral type G8 and the other a WD. Adopting a distance of 160 pc to Abell 35, Gatti et al. (1998) derived a projected separation between the

G-star and the WD of 18±5 a.u. The period of the system, for an assumed total mass of ~3 $M_\odot$, is more than 40 year. The system could, therefore, be somewhat similar to Sirius. Note though that the total mass of the system for two stars as described above should be closer to 2 $M_\odot$.

Burleigh (1999) reviewed the status of the hot WDs in Sirius and Procyon-type binary systems that were discovered from the *EUVE* and *ROSAT* FUV all-sky surveys. He mentioned the discovery of unresolved Sirius-type binaries as serendipitous results from UV and FUV observations, where the presence of a second component, much hotter than the optical primary, could be detected. The sky surveys performed by the *ROSAT* satellite with the Wide-Field Camera in the soft X-ray domain and in the FUV, and by *EUVE* only in the FUV, revealed more than 120 WDs, mostly isolated ones. In order to be discovered by these instruments, a WD had to be hotter than 2×10^4 K and be seen through no more than a few 10^{19} cm^{-2} neutral Hydrogen atoms (low HI column density).

Burleigh (1999) listed 20 new binary systems similar to Sirius, which were identified from the *ROSAT, EUVE,* and *IUE* observations. Among these, he pointed out two systems that are very interesting, from astrophysical points of view. These are y Puppis=HR 2875 (B5Vp) and θ Hydra=HR 3665 (B9.5V, see also Burleigh & Barstow 1999), both having hot WD companions. This indicates that the progenitors of these WDs must have been high-mass main-sequence stars, thus there may be even more Sirius-like systems to be found.

An even more extreme case could be λ Scorpius (Berghöfer et al. 2000). This is a B1.5IV star, but it shows an unusual soft X-ray excess in *Einstein* and *ROSAT* observations. The star has been shown to be a spectroscopic binary with a period of 5.959 days and a slightly eccentric (ϵ=0.29) orbit. In order to determine the nature of this system, Berghöfer et al. observed it with *EUVE* and showed that the most likely explanation for the soft X-ray/FUV excess is the presence of a hot WD companion. The apparent detection of a partial eclipse of the WD in the *EUVE* data, allowed Berghöfer et al. to establish its mass as ≥ 1.25 $M_\odot$ and its $T_{eff} \geq 64,000$ K. The system, therefore, is young and has not yet had time to circularize its orbit. This also means that after the red giant phase of the present-day WD, the system was left with an elliptical orbit, perhaps similar to the original orbit or perhaps caused by asymmetric mass loss from the system.

The observational results presented above indicate that there is no shortage of binary systems that harbor rather massive WDs along with fairly massive primaries. There seems to be a lack of such systems among the long-period binaries, but this could conceivably be the result of an

observational bias. One can expect to identify more such systems by corre-
lating UV or FUV data with colors in the optical domain and by searching
for normal stars with UV (or FUV) excesses. These could then be attributed
to hidden WD companions.

10.2. Sirius-basic data

In order to present a coherent, modern picture of the Sirius system I
collected below some of the relevant published measurements of the two
companions.

The data concerning Sirius B show strong fluctuations among the different values, because in many cases these are the results of model fitting
and the models are anchored to data of limited accuracy. The various values for T_{eff} fitted to Sirius B at different times changed widely in the
second half of the twentieth century, from close to 30,000 K to less than
23,000 K. Some values are listed in the *Searchable Index of White Dwarf
Observations*[1]. This reference mentions also an unpublished preprint by
Van Altena, from 1998, giving a new parallax value from the Yale Observatory. Sirius B is listed there with the following photometric values: V=8.44,
B–V=–0.3, and U–B=–1.04. These are essentially the values given by Rakos
& Havlen (1977), except for the B–V index. The system parallax is given
as 0".3816±0".002. Holberg's book (Holberg 2007) lists slightly different
effective temperature values with T_{eff}=25,193±37 K; this may originate
from partly unpublished data. The surface gravity of Sirius B also showed
fluctuations with time since the first attempts to measure it were published;
the recent values concentrate at log $g \simeq$8.6.

The collection of observational data presented here in Table 10.1, and
the extensive discussions on the evolution of individual stars and stars in
binary systems presented in the previous chapter, can be summarized by
two brief questions about the α CMa system that still do not have clear-cut
answers in the astronomical literature:

1. Is there a good astrophysical explanation of the Sirius system, as
   observed today?
2. In there a credible physical explanation for the reports of Sirius being
   red about 2,000 years ago?

I will approach these questions by first presenting a summary of the
observational parameters in Table 10.1 (with references for each measured
parameter in square brackets), then by describing briefly specific models
for the Sirius system.

---

[1]http://procyon.lpl.arizona.edu/WD/.

TABLE 10.1. Summary information for the Sirius stars

Physical parameter	Sirius A	Sirius B
Position	α=06:45:08.9173	δ=−16:42:58.017 [17]
Proper motion	μ_α =−0".5461 yr^{-1}	μ_δ =−1".2231 yr^{-1} [17]
Parallax	380.023±1.283 mas	380.023±1.283 mas [1, 9, 17]
Distance	2.631±0.009 pc	2.631±0.009 pc [1, 17]
Radial velocity	−8.7 km s^{-1}	−8.7 km s^{-1} [1, 16]
LISM N(HI) (cm^{-2})	(1.1–5)×10^{18} [10]	(1.1–5)×10^{18} [10]
Mass (M$_\odot$)	2.02±0.03 [16]	1.00±0.018 [1, 16]
Radius (R$_\odot$)	1.712±0.00025 [16]	(8.4±0.25) × 10^{-3}[8, 9]
T$_{eff}$ [K]	10500 [16]	25,193±37 [1, 6, 9, 16]
V	−1.42±0.03 [2]	8.44±0.03 [3]
B–V	+0.02 [13]	−0.03±0.03 [3]
U–B	0.00 [13]	−1.04±0.03 [3]
Rotation ($v \sin i$, km s^{-1})	16±1 [7]	0–600 [14, 15]
Carbon abundance (Sirius/Vega)	0.25× [4]	
Nitrogen abundance (Sirius/Vega)	1.66× [4]	
Oxygen abundance (Sirius/Vega)	0.54× [4]	
Iron abundance (Sirius/Sun)	5.0× [12]	
Magnetic field (gauss)	$\leq$ a few tens [5]	
Gravitational redshift (km s^{-1})		80.42±4.83 [16]
log g	4.3±0.1 [11]	8.556±0.01 [6, 9, 16]
Orbital period	50.075±0.103 [16]	
Epoch	1994.352±0.037 [16]	
a	7".50±0".03 [16]	
e	0.59657±0.0013 [16]	
i	138°.4376 ± 0°.298 [16]	
Ω	150°.334±0°.348 [16]	
ω	45°.685±0°.240 [16]	

References:

[1] Gatewood & Gatewood (1978)

[2] Stebbins (1950)

[3] Rakos & Havlen (1977)

[4] Lambert et al. (1982)

[5] Borra (1975)

[6] Greenstein et al. (1971)

[7] Kurucz et al. (1977)

[8] Böhm-Vitense et al. (1979)

[9] Holberg et al. (1998)

[10] Code et al. (1976) and Davis & Tango (1996)

[11] Frisch (1995)

[12] Bell & Dreiling (1981)

[13] Cayrel de Strobel et al. (1992)

[14] Eggen (1965)

[15] Thejll & Shipman (1986)

[16] Holberg

[17] Perrymam et al. (1997)

10.3. Synthesis of information

The specific and more detailed questions that different researchers attempted
to solve, regarding the Sirius system, were (a) what could the progenitor
star have looked like when Sirius B was on the main sequence, (b) how
exactly did the system evolve at the point where Sirius B left the main
sequence, (c) did any mass transfer between the WD progenitor and its
companion take place in the Sirius binary, (d) how could the present pattern of metallicity abundances in Sirius A be explained, (e) is the orbital
eccentricity original, or has it developed in the course of the off main-
sequence evolution of Sirius B, and (f) is there a reasonable explanation
for Sirius to have been perceived as red in antiquity. A number of models
specific to Sirius have been calculated. These were of increasing degrees
of complexity, but none was found to be fully satisfactory, perhaps due to
the simplifying assumptions adopted by their authors. Yet these models,
specifically the more recent ones, hint on what the true answer to these
questions might be.

10.3.1. Specific models for Sirius

The evolution of a binary system with a total mass of 7 $M_\odot$ was studied by
Lauterborn (1970). This is relevant for the case of Sirius, since some mass
estimates for the Sirius B progenitor were of this magnitude, and is briefly
discussed below. First, I summarize the calculations for the general case.
Then I bring Lauterborn's special remarks regarding the case of Sirius.
    The initial point chosen by Lauterborn (1970) for his model is of two
stars, one of 5 $M_\odot$ and the other of 2 $M_\odot$, at an initial distance of 301.9
$R_\odot$. At this distance, and with such component masses, the orbital period
is 227 days. Lauterborn allowed mass exchange between the two stars at
the end of central Helium burning, when Helium shell burning begins in
the more massive component. This mode of stellar evolution in a binary
system has been classified previously as *case C* by Kippenhahn & Weigert
(1967).
    The assumption adopted by Lauterborn (1970) is that the evolution of
the primary star (the predecessor of the white dwarf) can be considered as
if the star would be single. The secondary (nowadays the system primary,
Sirius A) is important only in determining the size of the Roche lobe. The
evolutionary calculation checked the radius of the evolving star and compared it to the Roche lobe size; when the stellar radius exceeded this size
the calculation removed some mass from the star to keep its size approximately equal to that of the Roche lobe. The initial 5 $M_\odot$ star was assumed
to have a chemical composition X=0.602, Y=0.354, and Z=0.044.

The 5 $M_\odot$ star evolved through the core Hydrogen and Helium burning phases without inflating beyond the Roche lobe, vindicating the initial assumption of evolution in isolation. After 7.5×10^7 yrs, with a 0.5 $M_\odot$ core of Carbon–Oxygen, the radius of the star reached the Roche lobe and during a rather short period of $(1.6–2.0) \times 10^5$ yrs almost 4 $M_\odot$ of stellar material were transferred to the companion. This, then, became the primary star in the system and the Roche lobes re-arranged themselves to account for this change in masses. Further description of the behavior of the model would not be applicable for the case of Sirius, as the present status of this system corresponds only to Lauterborn's post-phase-IIa (a very short phase of rapid mass loss following the red giant stage of the primary star and before the evolved star fills its Roche lobe).

After describing a number of binary systems that seem to support his theory, Lauterborn (1970) discussed specifically the Sirius system. His calculations, which included mass transfer between the components, could not show how a one solar mass white dwarf could be produced in a system with a separation of several 10^3 $R_\odot$. However, Lauterborn mentioned that the *Case C* mass exchange in a binary system, when the more massive star completed central Helium burning and burns Helium in a shell, could produce such a heavy WD. The problem seems to be the eccentricity exhibited by the present-day Sirius system ($e \approx 0.6$). For a system like Sirius, but where the orbits are circular, the critical Roche lobe has a size of 540 $R_\odot$. Mass transfer requires that the primary will reach a radius larger than this value during its evolution, but the evolutionary models known to Lauterborn did not reach such large radii (at most, 380 $R_\odot$ for a 5 $M_\odot$ star).

Roark (1977) studied the problem of binary-star evolution in which white dwarfs are formed. The particular systems that he considered are Sirius, Procyon, 40 Eridani, and Stein 2051. Roark addressed specifically the issue of mass loss from the WD progenitor star, assuming this to take place isotropically and without mass transfer. As a result, the orbital elements of the binary changed. Roark found that, for a precursor mass of 8.2 $M_\odot$, the likelihood of mass transfer through the inner Lagrangian point became negligible. By assuming that the metal enrichment of Sirius A is a result of some mass transfer through the L_1 point, Roark could even put a lower limit to the mass of the WD precursor star: 8 $M_\odot$, since otherwise mass transfer would not have taken place.

We saw above that stars of spectral class A, such as Sirius A, rotate relatively fast. However, the primary of the α CMa system is a very slow rotator. One wonders what was the role of magnetic braking during the binary evolution, if it happened at all, and whether it was linked with the metal-enrichment phase. Perhaps this was the mechanism that caused the braking of rotation in both A and B components. However, if there was a magnetic field that braked the rotation, where is it now?

Within different evolutionary models that could have driven Sirius B to look red in ancient times, the possibility of a planetary nebula-like ejection in historical times seems remote these days. Yet in the late-1970s this was not the case, the effective temperature of the WD was not accurately known, and it was possible that Sirius B could be a hot, young, bare stellar core. Brosch & Nevo (1978) searched unsuccessfully for traces of emission nebulosity around Sirius and managed to set an upper limit of 50 erg s^{-1} cm^{-2} ster^{-1} for any possible Balmer Hβ emission from the neighborhood of Sirius. A similar search, based on inspections of the Palomar Sky Survey prints and using near-IR observations, was reported by Mutz & Wyckoff (1992). Though they did not detect a nebula, they suggested that some intervening material near to the line of sight to Sirius might cause some reddening of the stars seen near α CMa.

A special treatment of the Sirius system, from an evolutionary point of view was by D'Antona (1982). The initial goal of this work was to examine the two outstanding peculiarities of the Sirius system: the high mass of the WD companion and the enhanced metallicity in the atmosphere of the primary. In these contexts, D'Antona included also the possibility of mass transfer from the WD precursor of Sirius B to the present-day primary. This is somewhat problematic, because the usual tidal mass transfer, as encountered in close binaries, would have circularized the orbit on a short time scale and the common-envelope evolution would have resulted in a cataclysmic binary, not in a detached system as observed now.

The evolutionary model produced by D'Antona (1982) assumed that tidal mass exchange did not take place in the Sirius system. The starting point was an estimate of the radius of Sirius B at the time it was a red giant. From various theoretical papers describing the evolution of intermediate-mass stars, D'Antona concluded that the radius of a red giant whose core is 1.05 M$_\odot$ should have been 600–800 R$_\odot$. An approximation to the closest distance between the red giant and Sirius A was obtained by estimating the size of the Roche lobe at periastron passage (from Paczyński 1976):

$$R_{RL} = R_p \left(0.38 + 0.2 \log \frac{M_B}{M_A}\right) \qquad (10.1)$$

Here R_p is the periastron distance and $R_p = a(1 - e)$, where a is the semi-major axis and e is the orbit eccentricity. For a 2.5 M$_\odot$ red giant progenitor of the WD, the Roche lobe size at periastron passage would already be significantly smaller than the red giant's diameter. The larger the red giant's mass, the larger its diameter would be.

D'Antona (1982) then required that the prompt planetary nebula ejection should modify the system eccentricity from ∼0.0 to ∼0.6, as observed today. This implies that the red giant should have been heavier than 3.0

$M_\odot$, but its mass should have been also ≤ 4.5 $M_\odot$ (or ≤ 5.25 $M_\odot$, if η=1.2)[2] because otherwise tidal mass transfer[3] would have taken place during the red giant phase.

Under these conditions, the progenitor system of Sirius should have contained a primary of ~ 4 $M_\odot$ (which is the present-day Sirius B) and a secondary of ~ 2.14 $M_\odot$ orbiting each other at a separation of ~ 6–7 a.u. and with a period of order 6–8 yrs. About 10^5 yrs after the initiation of double-shell burning in the primary circularization of the orbit took place and the eccentricity of the system virtually disappeared (e$\approx$0). At the moment of the atmosphere ejection into a planetary nebula the orbital period was ~ 10 yrs and Sirius B was almost filling its Roche lobe. This mechanism seems likely to produce the Sirius system as observed today.

A reasonable argument for accepting this evolutionary scheme was offered by the time scales of the problem. Since Sirius was assumed by D'Antona to belong to the Ursa Major stream (or the Sirius group), whose age was estimated from dynamical arguments at $\sim 3 \times 10^8$ yrs, one has to fit in this time slot both the evolution of a main-sequence star of the right mass to be the predecessor of Sirius B and the cooling of a stellar core of the proper mass to the present-day temperature of Sirius B. D'Antona (1982) adopted T_{eff}=26,000$\pm$1,000 K from Böhm-Vitense et al. (1979), and a mass of 1.0 $M_\odot$, to find an age of 1–1.5$\times 10^8$ yrs (after Lamb & Van Horn 1975) for such a WD. The evolutionary time left since the formation of the Sirius moving group is, therefore, matching the main-sequence lifetime of a 3–4 $M_\odot$ star, depending on its Helium content.

This model, however, failed to explain the anomalous chemical abundance in the atmosphere of Sirius A. At no point in the evolution there is a process capable of turning the secondary (the present-day Sirius A) into a metallic-line star with an enhanced abundance of metals, as observed. This indicated that the evolution must be further studied, to clarify which factor could be responsible for this enhancement. Note also that in a later paper (D'Antona & Mazzitelli 1995) the mass of the WD progenitor was set at a configuration (at the beginning of the asymptotic giant branch) of 7.0 $M_\odot$. The present slow rotation of Sirius A was also not explained by these models.

Joss et al. (1987) discussed the evolution producing a massive WD using empirical fits to model predictions. After dealing with single WD progenitor

---

[2]The parameter η is used in describing the mass loss rate in the giant phase: $dM/dt = 4 \times 10^{-13} \eta\, L/gR$ (Reimers 1975).

[3]Tidal mass transfer takes place because the tidal force from one of the stars distorts the other, which is in the process of expanding to a red giant configuration and fills its Roche lobe. Further evolution, in the presence of a companion star in a binary system, causes mass to flow from the more massive (and evolved) star to the other through the inner Lagrangian point, as was explained above.

stars, they extended the discussion to binary systems. Joss et al. assumed the initial system to be composed of a main-sequence secondary star and of a giant primary, which is losing its envelope through critical lobe overflow at periastron. Joss et al. confirmed that the conditions for stable overflow to produce eccentric orbits are unknown. Yet they assumed stable overflow if this happened in small increments at or near periastron, with the primary relaxing toward a hydrostatic configuration after periastron passage.

If the mass-loss process is stable, and if the system does not reach a common-envelope situation, the scenario of Joss et al. (1987) becomes relevant for the present-day Sirius system. This is because after the envelope of the giant star is lost, the stellar mass is reduced to nearly that of the core, m_c. Using a relation derived by Avni (1976) for the effective potential of the matter in the outer envelope of a star in a binary system, Joss et al. solved for the angular speed of the binary star:

$$\omega_k^2 = \frac{G\,M_{tot}\,(1-e)^3\,F^3(q, \xi_{min})}{R_{C,min}^3} \qquad (10.2)$$

Here ω_k is the system's angular rotation frequency, e is its eccentricity, F is the ratio of the Roche lobe size R_C to D, the separation of the binary components, q is the mass ratio of the two components, and $\xi = (\frac{D}{a})^3 \times (\frac{\Omega}{\omega_K})^2$ where Ω is the rotational angular momentum of the star around itself. For the Roche geometry, $e=0$ and $\Omega=\omega_k$, i.e., the orbit has circularized and the spin and orbital rotations are locked.

Equation (10.2) allowed Joss et al. (1987) to write the orbital period (in days) as:

$$P_{orb} \approx 0.116\,(\frac{R}{R_\odot})^{3/2}\,(\frac{m_c}{M_\odot})^{-1/2}\,(1-e)^{-3/2}\,[(1+q)^{1/3}\,F(q, \xi_{min})]^{-3/2}$$
$$(10.3)$$

where m_c is the core mass. From this, they derived an equation linking the eccentricity and the orbital period of a binary to the mass of the WD secondary:

$$(1-e)^{3/2}\,P_{orb} \approx 0.37\,(\frac{m_c}{M_\odot})^{-1/2}\,[\frac{R(m_c)}{R_\odot}]^{3/2} days \qquad (10.4)$$

or:

$$(1-e)^{3/2}\,P_{orb} \approx 8.4 \times 10^4\,(\frac{M}{M_\odot})^{11/2}\,[1 + (\frac{M}{M_\odot})^3 + 1.75\,(\frac{M}{M_\odot})^4]^{-3/2} days$$
$$(10.5)$$

The theoretical prediction could be tested against observational values with data from six well-studied binary systems, one of which was Sirius. Figure 2

of Joss et al. (1987) shows that Sirius is fairly close to the theoretical relation, supporting the idea of critical lobe overflow from the WD predecessor
onto the present Sirius A component of the system.

Joss et al. (1987) argued against an initially circular orbit for the WD
progenitor star and Sirius A because this would have led to a common-
envelope evolutionary path, with subsequent high mass loss and a large
reduction of the orbital period and eccentricity. If the mass loss occurred
via critical lobe overflow, this could have led to the present characteristics
of the Sirius system. However, their conclusion was that

> the hypothesis of critical lobe overflow by the giant progenitor of Sirius B
> is an intriguing but probably unsatisfactory explanation for the reported
> transition of Sirius [from a bright red star in historical times to the
> white one today] (my addition).

Any evolutionary model must also account for the present temperature
of the WD, i.e., it must fit the cooling of the WD in any time frame for the
entire system. For the timing of WD cooling we adopt Wood's WD cooling
data (Wood 1995)[4] where the cooling curve for a WD of 1.0 $M_\odot$, similar
in mass to Sirius B, is given. Such an object reaches a luminosity of 0.02
$L_\odot$ after $\sim 1.4 \times 10^8$ yr, at which time it has an effective temperature of
$\sim$24,000 K. The cooling behavior of 1.0 $M_\odot$, 0.9 $M_\odot$, and 0.8 $M_\odot$ WD models is shown in Figure 10.1. This depicts both the luminosity and effective
temperature as functions of time. The result for the possible cooling time
of Sirius B, given the most recent observational determinations, is approximately the same as when using the Kawaler (1999) cooling time formula.
The horizontal line corresponds to the present T_{eff} of Sirius B; it indicates
that the cooling age of the WD should be about 122 Myrs.

It seems that such scenarios, of a binary composed of an $\sim$8 $M_\odot$ star and
a $\sim$2 $M_\odot$ secondary formed about 3×10^8 yr ago as part of the Sirius moving
group, after most of the stars in the group already formed, could produce
under suitable conditions the Sirius system as observed today. The crucial
parameter in these scenarios is the requirement that mass transfer between
the giant stage of the primary star and the smaller object in the system
should take place only near periastron and that the ejection of the planetary nebula should be prompt and sufficient to produce the present-day
eccentricity from an almost-circular orbit. If these conditions are fulfilled,
it seems possible to produce a binary composed of a $\sim$2.1 $M_\odot$ primary and
a $\sim$1 $M_\odot$ Carbon–Oxygen WD.

The question of how Sirius A became an Am star is not fully solved
by these models, unless at the mass transfer through L_1 near periastron
passage, or from the accretion of part of the planetary nebula after its

---

[4]Available on-line at http://www.astro.fit.edu/wood/wd

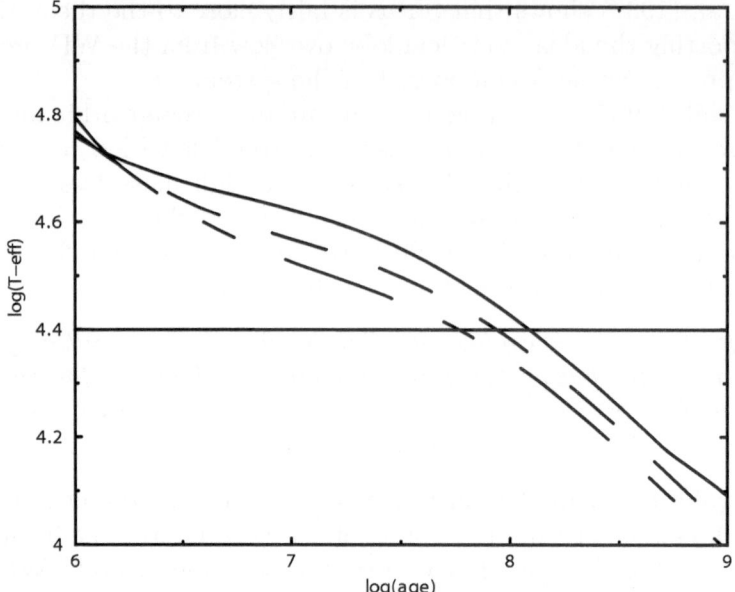

Figure 10.1.   Cooling behavior of three WDs following Wood (1995). The figure shows the effective temperature as a function of time in years for a 1.0 M$_\odot$ WD model (*solid line*), a 0.9 M$_\odot$ WD (*dashed line*), and 0.8 M$_\odot$ WD (*dot-dashed line*). The horizontal line corresponds to the present effective temperature of Sirius B. The 1 M$_\odot$ model reaches the present-day effective temperature of Sirius B after 122 Myrs.

ejection, the atmosphere of the secondary star (now Sirius A) received disproportionate amounts of CNO material (cf. Lambert et al. 1982). Another still unsolved question is why is Sirius A such a slow rotator in comparison with other A stars.

10.3.2. Red color in antiquity

The question of historical redness, discussed extensively in Section 3.1, may be solved in a number of ways using astrophysically-valid explanations. One of the first proposals, by D'Antona & Mazzitelli (1978), assumed that the Sirius system masqueraded as a red giant some 2,000 year ago when an event of unstable Hydrogen burning in a thin shell took place upon Sirius B. The Hydrogen was assumed to originate from the accretion of stellar wind material from the present system primary. A phase of Hydrogen burning phase lasting a few thousand years, when the luminosity of the system was enhanced and the WD appeared to observers as a red giant because of its extended envelope, could therefore provide a suitable explanation for historical redness.

The case of nuclear Hydrogen burning on a WD was studied by Shara et al. (1978), who concluded that such an event would form an EUV nova, as the extreme ultraviolet flux emitted then would reach a few 10^4 $L_\odot$. The thin envelope of the WD does not expand to form a quasi-red-giant envelope, but rather contracts and gives rise to a gradual increase in T_{eff}. This counters the proposal of D'Antona & Mazzitelli (1978). Note also that the accretion of a shell onto a WD followed by a classical nova explosion in historic times was proposed as the mechanism giving rise to the nebulosity around Z Cam, a recurrent dwarf nova (Shara et al. 2007).

Bonnet-Bidaud & Gry (1991) presented another scenario that could explain some of the reported properties of the Sirius system. According to their scenario, the historical evidence that Sirius was red could be explained by the transit of a small and dense cloudlet, about 0.01 $M_\odot$ and 0.02 pc in diameter, across the line of sight from the Sun to Sirius. As described above in Section 7.1, the LISM contains features of extremely small scales, thus the presence in antiquity of a very small cloud in the line-of-sight to Sirius cannot be excluded though no such body was yet identified. For a column density of $N(H) \approx 5.4 \times 10^{21}$ atoms cm^{-2}, the reddening would be $E(B–V)=1.0$ mag (Gry & Bonnet-Bidaud 1990) while Sirius itself would be dimmed to $V \approx 1.5$ mag, remaining one of the relatively bright stars but not one of the brightest. If the cloudlet is considered motionless, the time required for Sirius to cross the globule with its measured space motion would be $\sim$1,000 years. A slightly larger cloudlet, or a suitable direction of the vector of the relative motion with respect to the Sun-Sirius line of sight, might prolong the red phase of Sirius, as viewed from the Earth, sufficiently to cover the historical phase of the records.

Alternatively, Bonnet-Bidaud & Gry (1991) proposed that temporary reddening of Sirius could have been produced by a tidal event involving a possible third component of the system in a very eccentric orbit. Such hierarchical triple systems are known in the local Solar neighborhood. The presence of a third component in the inner Sirius system could, in principle, be responsible for some mass ejection that could provide the reddening agents. Using photometry derived from their coronagraphic observations of stars very close to Sirius, Bonnet-Bidaud & Gry showed that two stars (stars 3 and 4 in Table 4.1) seem to be redder than M8 stars. Their projected distances from Sirius are 165 to 205 a.u. and the stars could be in orbits around the Sirius system with periods >2,000 years. Bonnet-Bidaud & Gry calculated interaction times of these possible third companions of order one century, when they reside in the inner parts of the system, and noted that the reddening effect due to tidally-injected matter in the system could be even longer than the interaction time.

However, their subsequent paper, where this possibility of a distant third companion was tested by searching for common-proper-motion objects near Sirius (Bonnet-Bidaud et al. 2000), failed to support the existence of a third body in the system from among the candidate stars. The only place where such a third component might hide today would be very close to Sirius A and not at large distances, otherwise the various HST observations could have detected it. The unobserved region is smaller than ∼1 arcsec around Sirius A, that is, less than ∼2.6 a.u. This is the size of a region where a third star in the Sirius system, or some planets, might still hide, according to the suggestion of Benest & Duvent (1995). However, note that Benest & Duvent required this star to be in an approximately circular orbit, but in order for it to produce observable effects such as reddening in antiquity, the orbit ought to be elliptical.

It may be possible, probably, to formulate conditions on the eccentricity of possible orbits that would have the third body reside in this "hidden" region at present. However, it seems hard to allow strong tidal interactions to happen between it and Sirius A so that sufficient matter may be ejected, even temporarily, which would redden the Sirius system enough to make it visibly red. This is because the orbital period of the third body needs to be ≥2,000 year to correspond to the historic evidence and not to be detectable by present-day studies, while the third star must also be protected from ablation and evaporation during the red giant phase of the progenitor star that formed Sirius B (cf. Livio & Soker 1984). For a much shorter orbital period, which would shelter the third body during the extended giant phase, one needs to keep the third body from interacting with Sirius B on a secular time scale.

A scenario such as described above, whereby a third body in a highly elliptical orbit around Sirius A in the Sirius system on a ≥2,000 year orbit might, when arriving at periastron, trigger a tidal interaction that could inject sufficient mass in the inner system to redden Sirius A so as to make it visibly noticeable as a bright red star, cannot be ruled out at present. However, its adoption raises the interesting possibility that any Sirius C component, being now very close to Sirius A, could reach the tidal inter-action stage at any moment. Thus, if the tidal interaction scenario with a third companion could be correct, we may witness another red phase of Sirius at any moment, since this has not yet happened in the recent past!

In the more distant future Sirius will join the ranks of the fainter stars as it speeds away from the solar neighborhood. In 4–5 Myrs it will become fainter that the naked-eye visibility limit. After about three Milky Way years, more than 650 Myrs in the future, Sirius A will become a red giant. This stage will be followed by the ejection of a planetary nebula and the formation of another WD.

Soker & Rappaport (2000) studied the formation of planetary nebulae in various binary star type. They showed that systems with semi-major axes between 5 and 30 a.u. are likely to form PNs that have a bipolar shape, with very narrow waists, and leave remnant binaries with semi-major axes between 10 and 100 a.u. The future evolution of the system, when Sirius A will leave the main-sequence and evolve toward the AGB stage, will probably cause the semi-major axis of the system to approximately double after the second PN ejection. The final stage of the α CMa system would consist therefore of two rather distant and not very exciting WDs.

A less likely by much more spectacular future would be if the massive WD (now Sirius B) would accrete sufficient matter, some 0.4 $M_\odot$, from the evolving A component or from the ISM. This would bring its mass over the Chandrasekhar limit and the WD would explode in a type Ia supernova event, with a peak magnitude of ~ -19.3. Given very accurate galactic dynamics, one could predict where Sirius would be at that time to check whether a type Ia supernova would endanger the life on Earth.

10.4. Conclusions

This description of various Sirius-specific models may have "muddled the waters" somewhat, just at the conclusion of this collection of observational and theoretical arguments. Anyone can select whichever model they particularly like, but none is really satisfactory. Reasonable hopes exist that an even more detailed model of binary evolution, specifically tailored for Sirius, will be calculated that will avoid some of the ad hoc assumptions of D'Antona (1982). This now would have the benefits of improved observational results obtained from space-based platforms.

If I have to express my own opinions at the conclusion of this monograph, I would choose a mix between D'Antona's (1982) and Joss et al. (1987) evolutionary models that would include mass ejection from the progenitor of Sirius A as a means of producing the observed eccentricity and of modifying the surface metal abundances observed on Sirius A. The ejection of a PN as a means to induce the eccentricity seems also to be favored by Holberg (2007). The latter point, of the enhanced metal abundances, needs further study by the model calculators, as more accurate and extended abundance studies (e.g., Qiu et al. 2001) become available.

The explanation for the historical redness that I find more acceptable (disregarding the lack of similar evidence from non-Mediterranean civilizations) is that of a passage of a small ISM cloudlet in the line of sight between us and Sirius. That fits well the LISM structure described by Frisch in her many papers and may also match the enhanced ^{10}Be abundance in polar ice as well as the interstellar dust influx into the Solar System. I find hard to believe a finely-tuned model of a third body in the system that would produce intra-system material capable of reddening Sirius significantly for a sizable time of tens of orbits of Sirius B around A. On the other hand, an episodic reddening due to a nuclear burning stage in a thin envelope around the WD cannot be ruled out.

I hope that this presentation of the accumulated knowledge about Sirius will trigger further detailed studies of this system. The Sun is the prototype star that must be fully understood if theories of stellar structure and evolution are to be validated. Similarly, the Sirius system is the prototype binary star where much astrophysical theory can be tested. This is because both stars in this system are relatively bright by being nearby, and also because a wealth of observational data has already been collected. Future instruments will allow even higher angular resolution observations of the binary and of Sirius A, and will extended the wavelength basis with new observations. These will serve to further increase the data base to confront theoretical predictions, if the astrophysicists would accept this challenge.

Acknowledgements

In writing this book I was greatly helped by the availability of on-line resources. I used sometimes the SIMBAD database, operated at CDS, Strasbourg, France. I was also a constant user of the NASA's Astrophysics Data System Bibliographic Services (ADS) to consult abstracts and articles, some that were more than 200 years old. I also used occasionally the Google Book Search facility. I am grateful to all the persons who made these tools possible.

The Space Telescope Science Institute hosted me during a sabbatical year in 1997 when this book was started, and the Physics Department of the Michigan Technological University hosted me for some months of my 2007 sabbatical, when this book was completed. Both institutions are thanked for their hospitality.

I am grateful to Eugene de Geus, former science editor of Kluwer, for accepting the idea that a book about Sirius might make sense among those published in the ASSL series, and to the former and present Springer editorial staff for reminding me time and again of my promise to deliver a manuscript and for seeing this monograph to publication.

I acknowledge the many people I spoke with on the subject of Sirius during these many years, since the questions concerning Sirius and its nature began to bother me. Their valuable time was spent in explaining to me intricacies of white dwarf evolution, or of the nature of interstellar dust grains. They are too numerous to mention individually. I note with special thanks the late Jerome Mayo Greenberg, Dina Prialnik, Marcella Contini, Mario Livio, Mike Shara, Itzhak Goldman, and others.

I thank the following individuals for correspondence regarding Sirius, and in some cases for material related to Sirius they provided: Virginia Trimble, Turan Tuman, Robert Ceragioli, James Evans, Brad Schaefer, Göran Johansson, Igal Patel, and Yoav Yair. Gotthard Richter helped me locate the book *Rätsel um Sirius* by Dieter Herrmann; I am grateful to him and to Prof. Dr. Dieter Herrmann for the gift of this book. Rob van Gent kindly performed new calculations of the path of Sirius among the stars and allowed the inclusion of his plots in this book.

Dr. Gerald Picus provided the nice color photographs of the Farnese Atlas; I am grateful for his permission to use these in my book. Similarly, John A. Blackwell allowed me to use his colorful spectrum of Sirius and Dr. Galen R. Frysinger allowed me to use his wonderful color pictures of the Dogon. David Malin and Akira Fujii allowed the use of their sky image showing Sirius and its neighborhood. Anurag Shevade provided a color-composite image of Sirius and its immediate neighborhood created by combining UK Schmidt sky survey scans; this is used for the cover of the book. Jimmy Westlake allowed me to use his Mt. Wilson image of Sirius A and B.

Drafts of this book or sections of this book were read and commented on by Mike Shara, Dina Prialnik, Virginia Trimble, Elia Leibowitz, Robert Nemiroff, and Brad Schaefer. Brad, in particular, was the trigger of a major revision that hopefully improved the presentation.

References

Abbe, C. (1867) MNRAS **28**, 2.
Abbot, C.G. (1924) ApJ **60**, 87.
Abetti, G. (1954) *The History of Astronomy*, London: Sidgwick and Jackson.
Adams, W.S. (1902) ApJ **15**, 214.
Adams, W.S. (1911) ApJ **33**, 64.
Adams, W.S. (1914) PASP **26**, 198.
Adams, W.S. (1924) Proc Nat Acad Sci **11**, 382.
Adams, W.S. (1925) Obs. **48**, 337.
Adelman, S.J., Flores, R.S.C., Jr. & Patel, V.J. (2000) IBVS No. 4984.
Aitken, R.G. (1964) *The Binary Stars*, New York: Dover Publications, Inc.
Alfaro, E., Cabrera-Cano, J. & Delgado, A.J. (1991) ApJ **378**, 106.
Allen, R.H. (1963) *Star-Names and Their Meaning*, New York: Dover Publications, Inc. (reprint of 1899 edition).
Asiain, R., Figueras, F., Torra, J. & Chen, B. (1999) A&A **341**, 427.
Auer, L.H. & Woolf, N.J. (1965) ApJ **142**, 182.
Aumann, H.H. (1985) PASP **97**, 885.
Aumann, H., Gillett, F., Beichman, C., de Jong, T., Houck, J., Low, F., Neugebauer, G., Walker, R. & Wesselius, P. (1984) ApJL **278**, L23.
Auwers, A. (1864) MNRAS **25**, 38.
Auwers, A. (1892) AN **129**, 185.
Aveni, A.F. (1996) in *Astronomy Before the Telescope* (C. Walker, ed.), New York: St. Martin Press, p. 269.
Aveni, A. (2002) *Empires of Time*, Boulder, Colo., USA: University Press of Colorado.
Avni, Y. (1976) ApJ **209**, 574.
Babcock, H.W. (1958a) ApJS **3**, 141.
Babcock, H.W. (1958b) ApJ **128**, 228.
Baggaley, W.J. (2004) Earth Moon Planets **95**, 197.
Baggaley, W.J. & Neslušan, L. (2002) A&A **382**, 1118.
Bailey, C. (1974) Bull Sch Orient Afr Stud **37**(part 3), 580.
Barker, T. (1760) Philos Trans (1683–1775), **51** (1759–1760), 498.
Barstow, M.A., Bond, H.E., Burleigh, M.R., & Holberg, J.B. (2001) MNRAS **322**, 891.
Barstow, M.A., Bond, H.E., Holberg, J.B., Burleigh, M.R. Hubeny, I. & Koester, D. (2005) MNRAS **362**, 1134.
Becker, F. (1923) *Sternatlas*, Berlin: Fred. Dümmlers Verlagsbuchhandelung.
Bell, R.A. & Dreiling, L.A. (1981) ApJ **248**, 1031.
Benest, D. (1989) Astron Astrophys **223**, 361.
Benest, D. & Duvent, J.L. (1995) Astron Astrophys **299**, 621.
Bentley, J. (1823) *A Historical View of the Hindu Astronomy*, Calcutta: Baptist Mission Press.
Bergeron, P., Saffer, R.A. & Liebert, J. (1992) ApJ **394**, 228.
Bergeron, P., Ruiz, M.-T., Leggett, S.K., Saumon, D. & Wesemael, F. (1994) ApJ **423**, 456.
Bergeron, P., Liebert, J. & Fulbright, M.S. (1995) ApJ **444**, 810.

Berghöfer, W., Vennes, S. & Dupuis, J. (2000) ApJ **538**, 854.

Bernacca, P.L. & Perinotto, M. (1970) Contrib. dell'Oss. Astr. dell'Università di Padova in Asiago, No. 239.

Bernacca, P.L. & Perinotto, M. (1971) Contrib. dell'Oss. Astr. dell'Università di Padova in Asiago, No. 250.

Bertin, P., Lamers, H.J.G.L.M., Vidal-Madjar, A., Ferlet, R. & Lallement, R. (1995) A&A **302**, 899.

Bessel, F.W. (1844) MNRAS **6**, 136.

Beuermann, K., Burwitz, V. & Rauch, T. (2006) A&A **458**, 541.

Bicknell, P. (1989) Observatory **109**, 58.

Blaauw, A. (1961) BAN **15** (no. 505).

Bohlin, R.C., Savage, B.D. & Drake, J.F. (1978) ApJ **224**, 132.

Böhm-Vitense, E., Dettmann, T. & Kapranidis, S. (1979) ApJ **232**, L189.

Bond, G. (1862) Astron Nachr **57**, 131.

Bonnet-Bidaud, J.M., Colas, F. & Lecacheux, J. (2000) A&A **360**, 991.

Bonnet-Bidaud, J.M. & Gry, C. (1991) A&A **252**, 193.

Bono, G., Caputo, F., Cassisi, S., Marconi, M., Piersanti, L. & Tornambé, A. (2000) ApJ **543**, 955.

Borra, E.F. (1975) ApJ **202**, 741.

Botley, C.M. (1959) J British Astron Assoc **69**, 177.

Boyarchuk, A.A. & Snow, T.P. (1978) ApJ **219**, 515.

Bragaglia, A., Renzini, A. & Bergeron, P. (1995) ApJ **443**, 735.

Brecher, K. (1976) BAAS **8**(no. 3), 450.

Brecher, K. (1979) "Sirius Enigmas", in *Astronomy of the Ancients* (K. Brecher & M. Feirtag, eds.), Cambridge, Mass.: The MIT Press, pp. 91–115.

Britton, J. & Walker, C. (1996) in *Astronomy Before the Telescope* (C. Walker, ed.), New York: St. Martin Press, p. 42.

Broadfoot, A.L. et al. (1977) Space Sci Rev **21**, 183.

Brosch, N. (1992) Q. J. R. Astron. Soc **33**, 27.

Brosch, N. & Hoffman, Y. (1999) MNRAS **305**, 241.

Brosch, N. & Nevo, I. (1978) Observatory **98**, 136.

Brown, P.L. (1971) *What Star is that?* London: Thames and Hudson.

Bruhweiler, F.C. & Kondo, Y. (1982) ApJ **259**, 232.

Bruhweiler, F.C. & Kondo, Y. (1983) ApJ **269**, 657.

Budger, E.A.W. (1920) *Egyptian Hieroglyphic Dictionary* (on-line available pages).

Burleigh, M. (1999) in *proceedings of the 11th European Workshop on White Dwarfs*, ASP Conference Series no. 169 (S.-E. Solheim & E.G. Meistas, eds.), San Francisco: Astronomical Society of the Pacific, p. 249.

Burleigh, M.R. & Barstow, M.A. (1999) A&A **341**, 795.

Burleigh, M.R., Barstow, M.A., Schenker, K.J., Sills, A.I., Wynn, G.A., Dobbie, P.D. & Good, S.A. (2001) MNRAS **327**, 1158.

Burnham, S.W. (1891) MNRAS **51**, 378.

Burnham, S.W. (1893) MNRAS **53**, 482.

Burnham, S.W. (1897) MNRAS **57**, 453.

Burnham, R. (1966) *Burnham's Celestial Handbook; An Observer's Guide to the Universe Beyond the Solar System. A Descriptive Catalog and Reference Handbook of Deep-Sky Wonders for the Observer, Student, Research Worker, Amateur or Professional Astronomer.* Flagstaff, Ariz.: Celestial Handbook Publications.

Calame-Griaule, G. (1991) Curr Anthropol **32**(no. 5), 575–577.

Campbell, W.W. (1905) Pub Astron Soc Pacific **17**, 66.

Cannon, A.J. & Pickering, E.C. (1901) Ann Harvard Coll Obs **28**, 129.

Carruthers, G.R. (1968) ApJ **151**, 269.

Cash, W., Bowyer, S. & Lampton, M. (1978) ApJL **221**, L87.

Casperson, L.W. (1977) Appl Optics **16**(no. 12), 3183.

Castellani, V., Chieffi, A. & Straniero, O. (1992) ApJS **78**, 517.

Castro, S., Porto de Mello, G.F., & da Silva, L. (1999) MNRAS **305**, 693.

Cayrel de Strobel, G. (1981) Bull Inform CDS no. 20.

Cayrel de Strobel, G., Bentolila, C., Hauck, B. & Curchod, A. (1980) A&AS **41**, 405.

Cayrel de Strobel, G., Hauck, B., François, P., Thévenin, F., Friel, E., Mermilliod, M. & Borde, S. (1992) A&AS **95**, 273.

Ceragioli, R. (1992) Feruidus Ille Canis: the Lore and Poetry of the Dog Star in Antiquity, Ph.D. Thesis, Cambridge, Mass.: Harvard University.

Ceragioli, R.C. (1995) J History Astron **26**, 187.

Ceragioli, R.C. (1996) J History Astron **27**, 93.

Ceram, C.W. (1967) *Gods, Graves, and Scholars*, Toronto, New York, & London: Bantam Books.

Chacornac, M. (1864) MNRAS **25**, 40.

Chambers, G.F. (1912) *The Story of Eclipses*, New York: D. Appleby & Co.

Chandrasekhar, S. (1939) *An Introduction to the Study of Stellar Structure*, New York: Dover Publications (1957 republication of the original 1939 edition, with corrections).

Chapman-Rietschi, P.A.L. (1995) Q. J. R. Astr. Soc **36**, 337.

Chapman-Rietschi, P.A.L. (1997) J. R. Astr. Soc Can **91**, 133.

Chini, R., Krügel, E. & Kreysa, E. (1990) A&AL **227**, L5.

Cliver, E.W. (1989) Solar Phys **122**, 312.

Coblentz, W.W. (1922) ApJ **55**, 20.

Coblentz, W.W. (1927) Pop. Astron **35**, 137.

Code, A.D., Davis, J., Bless, R.C. & Hanbury-Brown, R. (1976) ApJ **203**, 417.

Code, A.D. & Meade, M. (1976) Wisconsin Ap., No. 30 (quoted in Bell & Dreiling 1981, op. cit.).

Comstock, G.C., (1897) ApJ **6**, 419.

Condos, T. & Reaves, G. (1971) PASP **83**, 834.

Conti, P.S. (1965) ApJ **142**, 1594.

Conti, P.S. (1970) PASP **82**, 781.

Crawford, I. & Price, R. (2000) Anglo-Aust Obs. Newsl (November), **95**, 6.

Czarny, J. & Felenbok, P. (1979) A&A **71**, 38.

D'Antona, F. (1982) A&A **114**, 289.

D'Antona, F. & Mazzitelli, I. (1978) Nature **275**, 726.

D'Antona, F. & Mazzitelli, I. (1987) in *The Second Conference on Faint Blue Stars*, IAU Colloquium No. 95 (A.G. Davis Philip, D.S. Hays & J.W. Liebert, eds.), New York: Schenectady, L. Davis Press, p. 635.

D'Antona, F. & Mazzitelli, I. (1990) Ann Rev A&A **28**, 139.

D'Antona, F. & Mazzitelli, I. (1995) in *White Dwarfs* (D. Koester & K. Werner, eds.), Berlin: Springer Verlag, p. 93.

Davis, J. & Tango, J. (1996) Nature **323**, 234.

Didelon, P. (1984) A&AS **55**, 69.

Dieter, N.H., Welch, W.J. & Romney, J.D. (1976) ApJ **206**, L113.

Dittrich, von E. (1928) Astron Nachr **231**, 385.

Dravins, D., Lindgren, L. & Torkelsson, U. (1990) A&A **237**, 137.

Dvorak, R. (1986) Astron Astrophys **167**, 379.

Dworetsky, M.M. (1974) ApJS **28**, 101.

Dziembowski, W., & Gesicki, K. (1983) Acta Astron. **33**, 183.

Eddington, A.S. (1924) MNRAS **84**, 308.

Eddy, J.A. (1974) Science **194**, 1035.

Eggen, O.J. (1950) ApJ **112**, 141.

Eggen, O.J. (1958) MNRAS **118**, 65.

Eggen, O.J. (1960) MNRAS **120**, 448.

Eggen, O.J. (1965) AJ **70**, 19.

Eggen, O.J. (1986) AJ **92**, 125.

Eggen, O.J. (1998) ApJ **116**, 782.

Eggen, O.J. & Greenstein, J.L. (1965) ApJ **141**, 83.

Eggleton, P.P. (1983) ApJ **268**, 368.

Ehrlich, A. (1959) *The Guide of our Skies* (in Hebrew: HaMadrich BeShmei Atrzeinu), Tel Aviv: Joshua Chechick Publisher, Ltd.

Evans, J. (1987) J Hist Astron **18**, 155 and 233.

Ferlet, R. (1999) A&A Rev **9**, 153.

Ferlet, R., Lallement, R. & Vidal-Madjar, A. (1986) A&A **163**, 204.

Ferrario, L., Wickramasinghe, D., Liebert, J. & Williams, K.A. (2005) MNRAS **361**, 1131.

Finsen, W.S. (1929) Circ Union Obs **80**, 87.

Flammarion, C. (1884) *Les Étoiles ét les Curiosités du Ciel*, Paris: Gauthier-Villars.

Fomenko, A.T., Kalashnikov, V.V. & Nosovsky, G.V. (1993) *Geometrical and Statistical Methods of Analysis of Star Configurations (dating Ptolemy's Almagest)*, Boca Raton: CRC Press.

Fontaine, G., Villeneuve, B. & Wilson, J. (1981) ApJ **243**, 550.

Fontaine, G., Brassard, P. & Bergeron, P. (2001) PASP **113**, 409.

Fowler, R.H. (1926) MNRAS **87**, 114.

Fox, P. (1925) Ann Dearborn Obs **2**, 115.

Frail, D.A., Weisberg, J.M., Cordes, J.M. & Mathers, C. (1994) ApJ **436**, 144.

Freire, R., Czarny, J., Felenbok, P. & Praderie, F. (1977) A&A **61**, 785.

Frisch, P.C. (1995) Space Sci Rev **72**, 499.

Frisch, P.C. (1997) *The Journey of the Sun*, Enrico Fermi Institute Report 97-23 (also astro-ph/9705231).

Frisch, P.C., Dorschner, J.M., Geiss, J., Greenberg, J. Mayo, Grün, E., Landgraf, M., Hoppe, P., Jones, A.P., Krätschmer, W., Linde, T.J., Morfill, G.E., Reach, W., Slavin, J.D., Svestka, J., Witt, A.N. & Zank, G.P. (1999) ApJ **525**, 492.

Frisch, P.C., & Slavin, J.D. (2003) ApJ **594**, 844.

Frisch, P.C. & York, D.G. (1983) ApJ **271**, L59.

Frisch, P.C. & York, D.G. (1986) in *The Galaxy and the Solar System* (R. Smoluchowsky, N. Bahcall & M.S. Matthews, eds.), Tucson: University of Arizona Press, p. 83.

Fusi-Pecci, F. & Renzini, A. (1976) A&A **46**, 447.

Gamow, G. (1940) *The Birth and the Death of the Sun*, New York: Mentor Books (9th edition 1960).

Gare, F. (1902) Observatory **25**, 166.

Gascoigne, S.C.B. (1950) MNRAS **110**, 15.

Gatewood, G.D. & Gatewood, C.V. (1978) ApJ **225**, 191.

Gatti, A.A., Drew, J.E., Oudmaijer, R.D., Marsh, T.R. & Lynas-Gray, A.E. (1998) MNRAS **301**, L33.

Geary, J.C. & Abt, H.A. (1970) AJ **75**, 718.

Gehlich, U.K. (1969) A&A **3**, 169.

Génova, R., Molaro, P., Vladilo, G. & Beckman, J.E. (1990) ApJ **355**, 150.

Giacconi, R. (1979) BAAS **11**, 408.

Gill, D., Sir (1898) MNRAS **58**, 81.

Girardi, L., Bertelli, G., Bressan, A., Chiosi, C., Groenewegen, M.A.T., Marigo, P., Salasnich, B. & Weiss, A. (2002) A&A **391**, 195.

Gliese, W.H. (1969) *Catalogue of Nearby Stars*, Veröffentl. Astron. Rechen-Inst. Heidelberg, no. 22.

Gliese, W.H. & Jahreiss, H. (1989) "Star Catalogues: A Centennial Tribute to A.N. Vyssotsky", in *The Third Catalogue of Nearby Stars. I. General View and Content* (A.G. Davis Philip & A.R. Upgren, eds.), New York: Schenectady, L. Davis Press, p. 3.

Gore, J.E. (1903) Obs **26**, 391.

Graff, K. (1931) Astron Nachr **241**, 143.

Gray, R.O. & Garrison, R.F. (1987) ApJS **65**, 581.

Greenstein, J.L. (1960) in *Stellar Atmospheres* (J.L. Greenstein, ed.), Chicago: University of Chicago Press, p. 687.

Greenstein, J.L. (1982) ApJ **258**, 661.
Greenstein, J.L., Oke, J.B. & Shipman, H.L. (1971) ApJ **169**, 563.
Greenstein, J.L., Boksenberg, A., Carswell, R. & Shortridge, K. (1977) ApJ **212**, 186.
Greenstein, J.L., Oke, J.B. & Shipman, H. (1985) Q Jl R Astr Soc **26**, 279.
Greenstein, J.L. & Trimble, V.L. (1967) ApJ **149**, 283.
Griffin, R.E.M., David, M. & Verschueren, W. (2000) A&AS **147**, 299.
Griffin, R. & Griffin, R. (1979) A&A **71**, 36.
Gry, C. & Bonnet-Bidaud, J.M. (1990) Nature **347**, 625.
Gry, C. & Jenkins, E.B. (2001) A&A **367**, 617.
Grzedzielski, S. & Lallement, R. (1996) Space Sci Rev **78**, 247.
Habing, H.J., Dominik, C., Jourdain de Muizon, M., Laureijs, R.J., Kessler, M.F., Leech, K., Metcalfe, L., Salama, A., Siebenmorgen, R., Trans, N. & Bouchet, P. (2001) A&A **365**, 545.
Hachisu, I., Kato, M. & Nomoto, K. (1999) ApJ **522**, 487.
Halley, E. (1717) Philos Trans Ser. I, **30**, 736.
Hamada, T. & Salpeter, E.E. (1961) ApJ **134**, 683.
Hanbury Brown, R., Davis, J., Lake, R.J.W. & Thompson, R.J. (1974) MNRAS **167**, 475.
Harpaz, A. (1995) *Stellar Evolution* (in Hebrew), Tel Aviv: Sifriat HaPoalim Publishing House, Ltd.
Harris, D.L., Strand, K. Aa. & Worley, C.E. (1963) in *Basic Astronomical Data* (K. Aa. Strand, ed.), Chicago: University of Chicago Press, p. 273.
Hartmann, L., Garrison, L.M. & Katz, A. (1975) ApJ **199**, 127.
Hawkes, R.L., & Woodworth, S.C. (1997) JRASC **91**, 218.
Hébrard, G., Mallouris, C., Ferlet, R., Koester, D., Lemoine, M., Vidal-Madjar, A. & York, D. (1999) Astron Astrophys **350**, 643.
Heiberg, J.L. (1898), *Syntaxis Mathematica of Claudius Ptolemy*, Leipzig 1903, digital version available on-line at Google Books.
Heiles, C. (1997) ApJ **481**, 193.
Henderson, T. (1839) MNRAS **5**, 5.
Henshaw, C. (1984) J British Astr Soc **94**(no. 5), 221.
Henyey, L.G., LeLevier, R. & Levée, R.D. (1955) PASP **67**, 341.
Herrmann, D.B. (1988) *Rätsel um Sirius* (Riddle of Sirius: 2nd edition), Berlin: Buchverlag Der Morgen.
Hetherington, N.S. (1980) Q Jl R Astr Soc **21**, 246.
Hetzler, C. (1935) ApJ **82**, 75.
Hill, G.M. & Landstreet, J.D. (1993) A&A **276**, 142.
Hills, J.G. & Dale, T.M. (1973) ApJ **185**, 937.
Hoffleit, D. & Jaschek, C. (1982) *The Bright Star Catalogue*, New Haven: Yale University Observatory (4th edition).
Holberg, J.B. (2007) *Sirius Brightest Diamond in the Night Sky*, Chichester, UK: Springer and Praxis.
Holberg, J.B., Wesemael, F. & Hubeny, I. (1984) ApJ **280**, 679.
Holberg, J.B., Barstow, M.A., Bruhweiler, F.C., Cruise, A.M. & Penny, A.J. (1998) ApJ **497**, 935.
Holberg, J.B., Barstow, M.A., Burleigh, M.R., Kruk, J.W., Hubeny, I. & Koester, D. (2004) Bull Am Astron Soc **36**, 1514.
Holden, E.S. (1896) AJ **17**, 26.
Holman, M.J. & Wiegert, P.A. (1999) Astron J **117**, 621.
Holmes, C.N. (1916) Pop. Astron **24**, 170.
Holweger, H., Hempel, M. & Kamp, I. (1999) A&A **350**, 603.
Huffer, C.M. & Whitford, A.E. (1934) PASP **46**, 221.
Huggins, W. (1877) The Observatory **1**, 4.
Hurwitz, M. et al. (1998) ApJ **500**, L1.
Hussey, W.J. (1896) PASP **8**, 183.

Iben, I. (1967) Ann Rev A&A **5**, 571.

Iben, I. (1991) ApJS **76**, 55.

Iben, I. & McDonald, J. (1986) ApJ **301**, 164.

Iben, I. & Tutukov, A.V. (1984) ApJS **54**, 335.

Izmodenov, V.V., Lallement, R. & Malama, Y.G. (1999) Astron Astrophys **342**, 13.

Jaschek, C. & Jaschek, M. (1987) *The Classification of Stars*, Cambridge: Cambridge University Press.

Jasinta, D.M.D. & Hidayat, B. (1999) A&AS **136**, 293.

Jenkins, L.F. (1963) *General Catalogue of Trigonometric Stellar Parallaxes*, New Haven: Yale University Observatory.

Johnson, H.M. (1961) Leaflet of the Astronomical Society of the Pacific, **8**, 255.

Johnson, H.L. (1966) Ann Rev A&A **4**, 193.

Johnson, H.L. (1980) Rev Mex A&A **5**, 25.

Jonckheere, R. (1918) MNRAS **78**, 657.

Jonckheere, R. (1930) Journal des Observateurs **13**, 76.

Jones, E.M. (1970) ApJ **159**, 101.

Joss, P.C., Rappaport, S. & Lewis, W. (1987) ApJ **319**, 180.

Kaler, J.B. (1996) *The Ever-Changing Sky, a Guide to the Celestial Sphere*, Cambridge: Cambridge University Press.

Kawaler, S.D. (1999) "11th European Workshop on White Dwarfs", edited by S.-E. Solheim and E.G. Meistas, Astr. Soc. Pacific Conference Series **169**, 158.

Keenan, P.C. & Hynek, J.A. (1950) ApJ **111**, 1.

Kervella, P., Thévenin, F., Morel, P., Bordé, P. & Di Folco, E. (2003) A&A **408**, 681.

Kidder, K., Holberg, J.B. & Wesemael, F. (1989) in *White Dwarfs* IAU Coll. 114 (G. Werner, ed.), Berlin, New York: Springer-Verlag, p. 350.

Kidder, K.M., Holberg, J.B. & Mason, P.A. (1991), AJ **101**, 579.

King, J.R., Villarreal, A.R., Soderblom, D.R., Gulliver, A.F. & Adelman, S.J. (2003) AJ 125, 1980.

Kippenhahn, R. & Weigert, A. (1967) Z Astrophys **45**, 251.

Knott, G. (1866) MNRAS **26**, 243.

Kocifaj, M. (1996) Contrib Astron Obs Skalnate Pleso **26**, 23.

Kodaira, K. (1967) Pub Astr Soc Japan **19**, 172.

Koester, D. (1979) A&A **72**, 376.

Koester, D. (1987) ApJ **322**, 852.

Koester, D., Dreizler, S., Weidemann, V. & Allard, N.F. (1998) A&A **338**, 612.

Kohl, K. (1964) Z Astrophys **60**, 115.

Kondo, Y., Talent, D.L., Barker, E.S., Dufour, R.J. & Modisette, J.L. (1978) ApJ **220**, L97.

Krüger, H., Landgraf, M., Altobelli, N. & Grün, E. (2007) Space Sci Rev, **109**, 401.

Krupp, E.C. (1983) *Echoes of the Ancient Skies; The Astronomy of Lost Civilizations*, New York: Harper & Row.

Kuchner, M.J. & Brown, M.E. (2000) PASP **112**, 827.

Kuiper G.P. (1932) BAN. **6**, 197.

Kuiper G.P. (1934) PASP **46**, 99.

Kurucz, R.L. & Furenlid, I. (1979) Smithsonian Astrophys Obs Special Report No. 387.

Kurucz, R.L., Peytremann, E. & Avrett, E.H. (1972) Smithsonian Astrophys Obs Special Report No. 209, p. 101.

Kurucz, R.L., Traub, W.A., Carleton, N.P. & Lester, J.B. (1977) ApJ **217**, 771.

Kwok, S. (2000) *The Origin and Evolution of Planetary Nebulae*, Cambrdige, UK: Cambridge University Press.

Lachaume, R., Dominik, C., Lanz, T. & Habing, H.J. (1999) A&A **348**, 879.

Lallement, R., Bertin, P., Vidal-Madjar, A. & Bertaux, J.L. (1994) Astron Astrophys **286**, 898.

Lamb, D.Q., Van Horn, H.M. (1975) ApJ **200**, 306.

Lambert, D.L., Roby, S.W. & Bell, R.A. (1982) ApJ **254**, 663.

Lampton, M., Tuohi, I., Garmire, G. & Charles, P. (1979) in *X-ray Astronomy, 21st Plenary Meeting of COSPAR* (W.A. Baity & L.E. Peterson, eds.), Elmsford, New York: Pergamon Press, p. 125.

Landsman, W.B., Murthy, J., Henry, R.C., Moos, H.W. & Linsky, J.L. (1984) ApJ **285**, 801.

Landsman, W., Simon, T. & Bergeron, P. (1995) in *White Dwarfs* (D. Koester & K. Werner, eds.), Berlin: Springer-Verlag, p. 191.

Latham, D. (1969) Smithsonian Astrophys Obs Special Report No. 321.

Lauroesch, J.T., Meyer, D.M., & Blades, J.C. (2000) ApJL **543**, L43.

Lauterborn, D. (1970) A&A **7**, 150.

Leroy, J.L. (1993a) A&AS **101**, 551.

Leroy, J.L. (1993b) A&A **274**, 203.

Leroy, J.L. (1999) A&A **346**, 955.

Li, J. & Wickramasinghe, T. (1998) MNRAS **300**, 718.

Liebert, J. (1980) Ann Rev A&A **18**, 363.

Liebert, J., Young, P.A., Arnett, D., Holberg, J.B. & Williams, K.A. (2005) ApJL **630**, L69.

Lindblad, B. (1922) ApJ **55**, 85.

Lindenblad, I.W. (1973) Astron J **78**, 205.

Linsky, J.L., Brown, A., Gayley, K., Diplas, A., Savage, B.D., Ayres, T.R., Landsman, W., Shore, S. & Heap, S.R. (1993) ApJ **402**, 694.

Linsky, J.L., Diplas, A., Ayres, T., Wood, B. & Brown, A. (1994) AAS **183**, ♯114.13.

Linsky, J.L., Diplas, A., Wood, B., Ayres, T. & Savage, B.D. (1995) ApJ **451**, 335.

Linsky, J., Redfield, S., Wood, B.E. & Piskunov, N. (2000) ApJ **528**, 756.

Lippincott, S.L. & Worth, M.D. (1966) Sky Telescope **31**, 4.

Livio, M. & Soker, N. (1984) MNRAS **208**, 763.

Lockyer, J.N. (1894) *The Dawn of Astronomy* (1992 edition by Kessinger Publishing), available digitaly at Google Books.

Lovi, G. (1989) Sky Telescope **77**(March), 287.

Lucke, P.B. (1978) A&A **64**, 367.

Luyten, W.J. (1922) PASP **34**, 365.

Lynn, W.T. (1887a) Observatory **10**, 194.

Lynn, W.T. (1887b) Observatory **10**, 104.

Lynn, W.T. (1902) Observatory **25**, 63.

Maffei, P. (1981) l'Astronomia **8**(January-February), 11.

Malek, J. (1994) Discussions in Egyptology **30**, 101.

Malin, D. & Murdin, P. (1984) *Colours of Stars*, Cambridge: Cambridge University Press.

Marshak, R.E. (1940) ApJ **92**, 321.

Marshak, R.E. & Blanch, G. (1946) ApJ **104**, 82.

Martin, C., Basri, G. & Lampton, M. (1982) ApJ **261**, L81.

Mazzitelli, I. & D'Antona (1987) in *The Second Conference on Faint Blue Stars*, IAU Colloquium No. 95 (A.G. Davis Philip, D.S. Hays & J.W. Liebert, eds.), New York: Schenectady L. Davis Press, p. 351.

McCluskey, S.C. (1987) Nature **325**, 87.

McCrea, W.H. (1972) Q Jl R Astr Soc **13**, 506.

McMahan, R.K. (1989) ApJ **336**, 409.

Mestel, L. (1952) MNRAS **112**, 583.

Mewe, R., Heise, J., Gronenschild, E.H.B.M., Brinkman, A.C., Schrijver, J. & den Boggende, A.J.F. (1975) ApJ **202**, L67.

Meyer, D.M. & Blades, J.C. (1996) ApJ **464**, L179.

Milliard, B., Pitois, M.L. & Praderie, F. (1977) A&A **54**, 869.

Minnaert, M.G.J. (1969) *Practical Work in Elementary Astronomy*, Dordrecht: D. Reidel Publishing company.

Moore, J.H. (1928) PASP **40**, 229.

Morgan, W.W. (1932) ApJ **75**, 46.

Munch, G. & Unsold, A. (1962) ApJ **135**, 711.

Mutz, S.B., & Wyckoff, S. (1992) Bull Am Astron Soc **24**, 1301.

Napiwotzki, R., Green, P.J. & Saffer, R.A. (1999) ApJ **517**, 399.

Nawar, S. (1999) Astrophys Sp Sci **262**, 477.

Norton, A.P. (1969) *A Star Atlas* (second reprint of the 15th edition, 1st edition published 1910), Edinburgh: Gall and Inglis.

Olano, C.A. (2001) AJ **121**, 295.

Oom, F. (1902) Observatory **25**, 167.

Oppolzer, T.H. Ritter v. (1887) *Canon der Finsternisse*, Wien: Kaiserlich-Königlich Hofund Staatsdruckerei.

Ostriker, J.P. & Hartwick, F.D.A. (1968) ApJ **153**, 797.

Paczyński, B. (1970) Acta Astron **20**, 47.

Paczyński, B. (1976) in *IAU Symp. 78, Structure and Evolution of Close Binary Systems* (B.E. Westerlund, ed.), Dordrecht, Holland: Reidel, p. 155.

Paerels, F.B.S., Bleeker, J.A.M., Brinkman, A.C. & Heise, J. (1988) ApJ **329**, 849.

Paerels, F.B.S. & Heise, J. (1989) ApJ **339**, 1000.

Pannekoek, A. (1961) *A History of Astronomy*, London: G. Allen & Unwin.

Paresce, F. (1984) AJ **89**, 1022.

Parish, P.W. (1985) J British Astr Soc **96**, 10.

Parthasarathy, M., Branch, D., Jeffery, D.J. & Baron, E. (2007) New Astron Rev **51**, 524.

Patterson, J.R., Moore, W.E. & Garmire, G.P. (1975) A&A **45**, 217.

Paul, G. (1979) Observatory **99**, 206.

Percy, J.R. & Wilson, J.B. (2000) PASP **112**, 846.

Perryman, M.A.C. et al. (1992) A&A **258**, 1.

Perryman, M.A.C. et al. (1997) A&A **323**, 49.

Pesch, P. & Pesch, R. (1977) Observatory **97**, 26.

Peters, C.A.F. (1851), Astron. Nach. **32**, 1.

Phillips, A.C. (1994) *The physics of stars*, Chichester, UK: John Wiley & Sons Ltd.

Pickering, E.C. (1890) Ann Harvard Coll Obs **27**, 1.

Plassmann, J. (1927) Himmelswelt **37**, 165.

Porchon, P. (1899) *Cours de Cosmographie*, Paris: Ancienne Librairie Germer Bailliére et Cie.

Praderie, F., Boesgaard, A.M., Milliard, B. & Pitois, M.L. (1977) ApJ **214**, 130.

Prialnik, D. (2000) *Stellar Structure and Evolution*, Cambridge: Cambridge University Press.

Price, R.J., Crawford, I.A. & Barlow, M.J. (2000) MNRAS **312**, L43.

Provencal, J.L., Shipman, H.L., Høg, E. & Thejll, P. (1998) ApJ **494**, 759.

Qiu, H.M., Zhao, G., Chen, Y.Q. & Li, Z.W. (2001) ApJ **548**, 953.

Quémerais, E., Sandel, B., Lallement, R. & Bertaux, J.L. (1995) A&A **299**, 249.

Rakos, K.D. (1974) A&A **34**, 157.

Rakos, K.D. & Havlen, R.J. (1977) A&A **61**, 185.

Ramella, M., Boehm, C., Gerbaldi, M. & Faraggiana, R. (1989) A&A **209**, 233.

Redfield, S. & Linsky, J.L. (2000) ApJ **534**, 825.

Reid, I. Neill (1996) AJ **111**, 2000.

Reimers, D. (1975) Mem Soc Sci Liége Series **8**, 369.

Richier, J., Michaud, G. & Turcotte, S. (2000) ApJ **529**, 338.

Ridpath, I. (1988) Observatory **108**, 130.

Riegler, G.R. & Garmire, G.P. (1975) A&A **45**, 213.

Roark, T.P. (1977) Rev Mex Astron Astrophys **3**, 113.

Roman, N.G. (1949) ApJ **110**, 205.

Ruiz, M., Leggett, S. & Allard, F. (1997) ApJ **491**, L107.

Sadakane, K. & Okyudo, M. (1989) Pub Astr Soc Japan **41**, 1055.

Sadakane, K. & Ueta, M. (1989) Pub Astr Soc Japan **41**, 279.

Safford, T.H. (1862) MNRAS **22**, 145.

Salaris, M., Domingues, I., García-Berro, E., Hernanz, M., Isern, J. & Mochkovitch, R. (1997) ApJ **486**, 413.

Savedoff, M.P., Van Horn, H.M., Wesemael, F., Auer, L.H., Snow, T.P. & York, D.G. (1976) ApJ **207**, L45.

Schaeberle, J.M. (1894a) The Observatory **17**, 304.

Schaeberle, J.M. (1894b) PASP **6**, 237.

Schaeberle, J.M. (1894c) AJ **14**, 46.

Schaefer, B.E. (1985) Sky Telescope **70**, 261.

Schaefer, B.E. (1991) PASP **103**, 645.

Schaefer, B.E. (1993) Vistas Astron **36**, 311.

Schaefer, B.E. (1998) Sky Telescope **95**, 57.

Schaefer, B.E. (2000) J Hist Astron **31**, 149.

Schaefer, B.E. (2005) J Hist Astron **36**, 167.

Schaller, G., Schaerer, D., Meynet, G. & Maeder, A. (1992) A&AS **96**, 269.

Schatzman, E. (1958) *White Dwarfs*, Amsterdam: North-Holland Publishing Co.

Schliemann, H. (1880) *Ilios: The city and the Country of the Trojans*, London: John Murray, (available on-line through Google Books).

Schlosser, W. & Bergmann, W. (1985) Nature **318**, 45.

Schlosser, W. & Bergmann, W. (1987) Nature **325**, 89.

Schoch, C. (1924) MNRAS **84**, 731.

Schönberner, D. & Harmanec, P. (1995) A&A **294**, 509.

Schroeder, D.J., Golimowski, D.A., Brukard, R.A., Burrows, C.J., Caldwell, J.J., Fastie, W.G., Ford, H.C., Hesman, B., Kletskin, I., Krist, J.E., Royle, P. & Zubrowski, R.A. (2000) Astron J **119**, 906.

Schwarzschild, M. (1958) *Structure and Evolution of the Stars*, New York: Dover Publications, Inc (1965 republication of the 1958 edition).

Secchi, A. (1863) Astron Nachr **59**, 71.

Secchi, A. (1868) MNRAS **28**, 196.

See, T.J.J. (1927) Astron Nachr **229**, 244.

Sekiguchi, M. & Fukugita, M. (2000), AJ **120**, 1072.

Serviss, G.P. (1908) *Astronomy with the Naked Eye*, New York and London: Harper & Brothers Publishers.

Severny, A. (1970) ApJ **159**, L73.

Shallis, M.J. & Blackwell, D.E. (1980) A&A **81**, 336.

Shara, M.M., Prialnik, D. & Shaviv, G. (1978) A&A **61**, 363.

Shara, M.M. et al. (2007) Nature **446**, 159.

Shipman, H.L., Margon, B., Bowyer, S., Lampton, M., Paresce, F. & Stern, R. (1977) ApJL **213**, L25.

Sion, E.M., Greenstein, J.L., Landstreet, J.D., Liebert, J., Shipman, H.L. & Wegner, G.A. (1983) AJ **269**, 253.

Slettebak, A. (1954) ApJ **119**, 146.

Slettebak, A. (1955) ApJ **121**, 653.

Smith, M.A. (1976) ApJ **203**, 603.

Snedegar, K.V. (1997) Mercury **26** (6 November–December), 12.

Soderblom, D.R. & Mayor, M. (1993) AJ **105**, 226.

Soker, N. & Rappaport, S. (2000) ApJ **538**, 241.

Sonett, C., Morfill, G. & Jokipii, J. (1987) Nature **330**, 458.

Song, I., Caillault, J.-P., Barrado y Navascués, D. & Stauffer, J.R. (2001) ApJ **546**, 352.

Sowell, J.R., Trippe, M., Caballero-Nieves, S.M., & Houk, N. 2007, AJ **134**, 1089.

Spear, G.G., Kondo, Y. & Henize, K.G. (1974) ApJ **192**, 615.

Staal, J.D.W. (1984) *Stars of Jade*, Decatur (GA): Writ Press.

Stebbins, J. (1950) MNRAS **110**, 416.

Stecher, T.P. (1970) ApJ **159**, 543.

Steffey, P.C. (1982) Sky Telescope **84** (no. 3, September), 266.

Stenzel, A. (1928) Astron Nachr. **231**, 387.

Stott, C. (1995) *Celestial Charts*, London: Studio Editions Ltd.

Strom, S.E., Gingerich, O. & Strom, K.M. (1966) ApJ **146**, 880.

Strömgren, G. (1926) PASP **38**, 44.

Struve, O. (1864) MNRAS **24**, 149.

Struve, O. (1866) MNRAS **26**, 268.

Struve, O. (1955) Sky Telescope **14**, 461.

Swings, J.-P. (1982) Perkin-Elmer Tech News **11**(no. 2), p. 22.

Takada-Hidai, M. & Jugaku, J. (1993) in *Peculiar Versus Normal Phenomena in A-Type and Related Stars* (M.M. Dworetsky, F. Castelli & R. Faraggiana, eds.), ASP Conference Series **44**, 310.

Talon, S., Richard, O. & Michaud, G. (2006) ApJ **645**, 634.

Tang, T.B. (1986) Nature **319**, 532.

Temple, R.K.G. (1975) Observatory **95**, 52.

Temple, R.K.G. (1981) Zetetic Scholar **8**, 29.

Temple, R.K.G. (1987) *The Sirius Mystery*, Rochester, Vt.: Destiny Books.

Thejll, P. & Shipman, H.L. (1986) PASP **98**, 922.

Thurston, H. (1994) *Early Astronomy*, New York: Springer-Verlag.

Tinbergen, J. (1982) A&A **105**, 53.

Titus, J. & Morgan, W.W. (1940) ApJ **92**, 256.

Tomkin, J. (1998) Sky Telescope **95**(April), 59.

Trimble, V. (1999) Bull Astr Soc India **27**, 549.

Trimble, V. & Greenstein, J.L. (1972) ApJ **177**, 441.

Tuman, V.S. (1983) Q J R Astr Soc **24**, 14.

Uesugi, A. & Fukuda, I. (1970) Contrib Inst Ap and Kwasan Obs., No. 189.

Umeda, H., Nomoto, K., Yamaoka, H. & Wanajo, S. (1999) ApJ **513**, 861.

Van Albada-Van Dien, E. (1977) A&AS **29**, 305.

van Beek, W.E.A. (1991) Curr Anthropol **32**(2, April), 139.

van de Kamp, P. (1961) PASP **73**, 389.

van de Kamp, P. (1969) PASP **81**, 5.

van de Kamp, P. & Barcus, L.F. (1936) AJ **45**, 124.

van den Bos, W.H. (1929) Circ Union Obs **80**, 68.

van den Bos, W.J. (1960) J d Obs **43**, 145.

Van Gent, R.H. (1984) Nature **312**, 302.

Van Gent, R.H. (1986) Sky Telescope (December), letter to the editor.

Van Gent, R.H. (1987) Nature **325**, 87.

Vassiliadis, E. & Wood, P.R. (1994) ApJS **92**, 125.

Vennes, S. (1999) ApJ **525**, 995.

Vennes, S., Christian, D. & Thorstensen, J.R. (1998) ApJ **502**, 763.

Vogel, H.C., (1895) ApJ **2**, 333.

Volet, Ch. (1931) Bul Astr **7**, 24.

Volet, Ch. (1932) Bull Astr **8**, 51.

Vyssotsky, (1933) ApJ **78**, 1.

Warner, B. (1996) in *Astronomy Before the Telescope* (C. Walker, ed.), New York: St. Martin Press, p. 304.

Warner, B. & Sneden, C. (1988) MNRAS **234**, 269.

Watson, J.K. & Meyer, D.M. (1996) ApJ **473**, L127.

Wegner, G. (1981) AJ **86**, 264.

Weidemann, V. (1977) A&A **59**, 411.

Weidemann, V. (1990) ARAA **28**, 103.

Weidemann, V. & Koester, D. (1983) A&A **121**, 77.

Weidemann, V. & Koester, D. (1984) A&A **132**, 195.

Wells, R.A. (1996) in *Astronomy Before the Telescope* (C. Walker, ed.), New York: St. Martin Press, p. 28.

Wesemael, F. (1985) Q Jl R Astr Soc **26**, 273.

Wesemael, F., Greenstein, J.L., Liebert, J., Lamontagne, R., Fontaine, G., Bergeron, P. & Glaspey, J.W. (1993) PASP **105**, 761.

West, M.L. (1971) The Classical Quarterly, New Series **21** (No. 2), 365.

Westgate, C. (1933) ApJ **78**, 48.

Whittet, D.C.B. (1999) MNRAS **310**, 355.

Wickramasinghe, D.T. & Ferrario, L. (2000) PASP **112**, 873.

Wiese, W.L. & Kelleher, D.E. (1971) ApJ **166**, L59.

Winget, D.E., Hansen, C.J., Liebert, J., Van Horn, H.M., Fontaine, G., Nather, R.E., Kepler, S.O. & Lamb, D.Q. (1987) ApJL **315**, L77.

Wonnacott, D., Kellet, B.J. & Stickland, D.J. (1993) MNRAS **262**, 277.

Wood, B.E., Müller, H.-R. & Zank, G.P. (2000) ApJ **542**, 493.

Wood, M.A. (1995) in *Proceeding of the 9th European Workshop on White Dwarfs* (D. Koester & K. Werner, eds.), Berlin: Springer, p. 41.

Young, P.A. & Arnett, D. (2005) ApJ **618**, 908.

Yushchenko A. et al. 2007, The Seventh Pacific Rim Conference on Stellar Astrophysics **362**, 46.

Zagar, F. (1932) Atti del R. Istituto Veneto di S.L.A. **91**, 1047.

Zank, G.P. & Frisch, P.C. (1999) ApJ **518**, 965.

Zhao, G., Qiu, H.M., Chen, Y.Q. & Li, Z.W. (2000) ApJ Suppl. **126**, 461.

Ziznovsky, J. & Zverko, J. (1995) Contrib Astron Obs Skalnate Pleso **25**, 39.

Zwiers, H.J. (1899) Astron Nachr **150**, 221.

Index

Abundance, 44, 103–107, 117, 123, 124, 127, 129, 135, 147, 151, 168–169, 183, 184, 189, 190, 193, 199
Aldebaran (α Tau), 37, 38
Almagest, 10, 36–37, 47
Am star, 103, 105, 107, 111, 121, 168–169, 183, 195
Antares (α Sco), 37, 38, 46
Arcturus (α Boo), 37, 38, 43
Astronomical Netherlands Satellite (ANS), 125–126
Atlas Farnese, 21–22

Babylon, 17–20, 29, 36, 41, 49
Bessel, F.W., 57–59, 66, 69, 71, 88
Betelgeuse (α Ori), 37, 38, 46, 47
Bighorn, 29–30

Canopus (α Car), 28, 31, 39, 56, 95
Capella (α Aur), 46, 137
Carbon core, 129, 162
Ceragioli, B.C 25, 35, 37–39, 44–45, 134, 201
Chandra, 126, 127, 152
Chandrasekhar, S., 155–161, 181, 199
CNO elements, 104–105, 150, 158, 169, 173, 196
Color-magnitude (CM))diagram, 52, 53, 54, 107, 170
Copernicus satellite, 107, 108, 123, 135

Denderah, 13, 15
Dieterlen, G., 61, 64, 67
Dogon, 6, 52, 60–69, 201

Einstein Observatory, 126
Extinction, 17, 30, 41, 48–52, 68, 94–96, 133, 134, 143, 144, 148
Extreme Ultraviolet Explorer (EUVE), 79, 124, 125, 127, 136, 138, 160, 186–187

Flammarion, C. 9, 26, 36–38, 47, 55, 56, 66, 72

Gemini 12, 121, 122
Griaule, M., 67–68

Halley, Sir, Edmund., 55–56, 92
Hathor, 12–15
Hatra, 41–42
Herzsprung-Russel (HR) diagram, 159, 165
Hesiod, 24
High Energy Astronomical Observatory (HEAO), 126–127
High Precision Parallax Collecting Satellite (Hipparcos), 93
Holberg, J., 76, 93, 110, 121, 123–125, 129, 161, 188–189, 199
Homerus, 20, 24
Hubble Space Telescope (HST), 3, 79, 84, 85, 88, 115–116, 128, 134, 136, 138–139, 143, 186, 198
Hyades, 170
Hydrogen flash, 172, 173, 175

215